Rethinking Mathematics

RETHINKING MATHEMATICS

Teaching Social Justice by the Numbers

Edited by
Eric Gutstein
and Bob Peterson

A RETHINKING SCHOOLS PUBLICATION

RETHINKING MATHEMATICS
Teaching Social Justice by the Numbers
Edited by Eric Gutstein and Bob Peterson

Dedicated to
Claudia Zaslavsky
1917–2006
Educator, Multiculturalist, Anti-Racist, Ethnomathematician

RETHINKING SCHOOLS, LTD., is a nonprofit educational publisher of books, booklets, and a quarterly journal on school reform, with a focus on issues of equity and social justice. To request additional copies of this book or a catalog of other publications, or to subscribe to the quarterly journal *Rethinking Schools,* contact:

Rethinking Schools
1001 East Keefe Avenue
Milwaukee, Wisconsin 53212 USA
800-669-4192
www.rethinkingschools.org

Rethinking Mathematics: Teaching Social Justice by the Numbers
© 2006, Rethinking Schools, Ltd.

Cover design, page design and layout by Joanna Dupuis, design@joannadupuis.com
Production Editors: Leon Lynn and Bob Peterson

Cartoons on pages 50 and 162 are by Fred Wright/UE NEWS. Used with the permission of UE NEWS/United Electrical, Radio and Machine Workers of America (UE).

All rights reserved. Except as permitted below, no part of this book may be reproduced in any form or by any means, including electronic, without the express prior written permission of the publisher, except for brief quotation in an acknowledged review. Individuals may photocopy excerpts from this book for use in educational settings (physical settings only; Internet use is not permitted), as long as such material is furnished free to students or workshop participants. For sale of any portion of this book as part of course packets, contact the Copyright Clearance Center for permissions and appropriate fees. If you have any questions, contact the publisher at the address above.

ISBN 0-942961-54-4

Contents

Acknowledgements

Putting together a book like this would not have been possible if it had not been for the support and work of many individuals. We would like to thank all those math teachers and authors we have worked with to include their writings and ideas in this book. We have learned from their teaching and writing as we hope the readers of this book will as well.

We would like to thank, in particular, Bill Bigelow, Rethinking Schools classroom editor, who worked closely with us giving advice and feedback on the book as a whole and many specific articles; Leon Lynn, Rethinking Schools book editor; and Catherine Capellaro, managing editor of *Rethinking Schools* magazine.

Eric Gutstein offers special thanks to the students and community of Rivera Elementary School, in Chicago, and to Pauline Lipman for her ever-critical support and analysis.

Bob Peterson thanks the fifth-grade students he has taught at La Escuela Fratney, in Milwaukee, many of whom have provided the inspiration and challenges that have helped him become the social justice teacher that he is. He also thanks the entire Fratney community and dedicated staff, who continue to inspire him. Finally, he thanks Barbara Miner for her support and feedback.

Rethinking Mathematics, like all Rethinking Schools books, was supported by the broader Rethinking Schools community, especially the editors, Wayne Au, Terry Burant, Linda Christensen, Stan Karp, Larry Miller, Kelley Dawson Salas, Rita Tenorio, and Stephanie Walters; and the editorial associates, David Levine and Kathy Swope; and the office staff, Mike Trokan and Susan Bates.

Introduction

I thought math was just a subject they implanted on us just because they felt like it, but now I realize that you could use math to defend your rights and realize the injustices around you.... [N]ow I think math is truly necessary and, I have to admit it, kinda cool. It's sort of like a pass you could use to try to make the world a better place.

— **Freida, ninth grade, Chicago Public Schools**

Some students would prefer to have a dentist drill their teeth than to sit through a math class. Others view math class as a necessary but evil part of getting through school. Still others enjoy playing and working with numbers and problems.

We agree with Freida. Math is often taught in ways divorced from the real world. The alternative we propose in this book is to teach math in a way that helps students more clearly understand their lives in relation to their surroundings, and to see math as a tool to help make the world more equal and just.

In a "rethought" math class, teachers make mathematics more lively, accessible, and personally meaningful for students, who in turn learn in more depth.

The articles in this book provide examples of how to weave social justice issues throughout the mathematics curriculum and how to integrate mathematics into other curricular areas. This approach seeks to deepen students' understanding of society and to prepare them to be critical, active participants in a democracy.

The elementary school, middle school, high school, and college teachers who have contributed to this book also note the many potential benefits of such a social justice approach to mathematics. Among them:

- Students can recognize the power of mathematics as an essential analytical tool to understand and potentially change the world, rather than merely regarding math as a collection of disconnected rules to be rotely memorized and regurgitated.
- Students can deepen their understanding of important social issues, such as racism and sexism, as well as ecology and social class.
- Students can connect math with their own cultural and community histories and can appreciate the contributions that various cultures and peoples have made to mathematics.
- Students can understand their own power as active citizens in building a democratic society and become equipped to play a more active role in that society.
- Students can become more motivated to learn important mathematics.

These benefits come both when teachers reshape the mathematics curriculum with a social justice vision and when they integrate social justice mathematics across the curriculum into other subjects, such as social studies, science, health, reading, and writing.

An Essential Tool for Understanding and Changing the World

To have more than a surface understanding of important social and political issues, mathematics is essential. Without mathematics, it is impossible to fully understand a government budget, the impact of a war, the meaning of a national debt, or the long-term effects of a proposal such as the privatization of Social Security. The same is true with other social, ecological, and cultural issues: You need mathematics to have a deep grasp of the influence of advertising on children; the level of pollutants in the water, air, and soil; and the dangers of the chemicals in the food we eat. Math helps students understand these issues, to see them in ways that are impossible without math; for example, by visually displaying data in graphs that otherwise might be incomprehensible or seemingly meaningless.

As an example, consider racial profiling. This issue only becomes meaningful when viewed through a mathematical lens, whether or not the "viewer" appreciates that she or he is using mathematics. That is, it is difficult to declare that racial profiling occurs unless there is a sufficiently large data set and a way to examine that data. If, for example, 30 percent of drivers in a given area are African Americans, and the police stop six African-American drivers and four white drivers, there is weak evidence that racial profiling exists. But if police stop 612 African-American drivers and 423 whites, then there is a much stronger case.

The explanation lies in mathematics: In an area where only 30 percent of the drivers are black, it is virtually impossible for almost 60 percent of more than 1,000 people stopped randomly by the police to be black.

The underlying mathematical ideas — (dis)proportionality, probability, randomness, sample size, and the law of large numbers (that over a sufficiently large data set, the results of a probability simulation or of real-world experiences should approximate the theoretical probabilities) — all become part of the context that students must understand to really see, and in turn demonstrate, that something is amiss. Thus with a large data set, one can assert that a real problem exists and further investigate racial profiling. For youth, racial profiling may mean being "picked on," but the subtleties and implications are only comprehensible when the mathematical ideas are there. (See "Driving While Black or Brown," page 16.)

When teachers weave social justice into the math curriculum and promote social justice math "across the curriculum," students' understanding of important social matters deepens. When teachers use data on sweatshop wages to teach accounting to high school students or multi-digit multiplication to upper-elementary students, students can learn math, but they can also learn something about the lives of people in various parts of the world and the relationship between the things we consume and their living conditions. (See "Sweatshop Accounting," page 53, and "Sweatshop Math," page 160.)

Moreover, to understand some issues, students need to combine math with other subjects. For example, although the government releases unemployment figures monthly, Marilyn Frankenstein explains in "Reading the World with Math" (page 19) that how unemployment figures are reported profoundly affects one's understanding of what really goes on in our communities. Frankenstein points out that it's easy enough to figure unemployment percentages, but who gets counted as unemployed and who is — or is not — considered part of the workforce are political questions. The unemployment rate changes depending on these decisions. Thus math needs social studies, and social studies needs math.

Connecting Math with Students' Cultural and Community Histories

Rethinking math also means using culturally relevant practices that build on the knowledge and experiences of students and their communities. Many of these approaches have been developed by teachers and then described and theorized by researchers of color, such as Gloria Ladson-Billings and William Tate. A guiding principle behind much of this work is that teachers should view students' home cultures and languages as strengths upon which to build, rather than as deficits for which to compensate. In "Race, Retrenchment, and the Reform of School Mathematics" (page 31), Tate offers the simple example of a teacher's failure to reach her students because she uses story problems that are not grounded in the students' culture;

while Luis Ortiz-Franco ("Chicanos Have Math in Their Blood," page 70) encourages teachers to teach about the base-20 Mayan number system as a way to emphasize, to both Chicano students and others, that math has deep roots in indigenous cultures in the Americas. David Levine's article about the Algebra Project started by Bob Moses (see "Radical Equations," page 90) provides another example of teachers connecting with students' cultural and community histories to promote deeper student learning. Moses summarized the importance of these connections in his book on the project:

> [I]n the Algebra Project we are using a version of experiential learning; it starts with where the children are, experiences that they share. We get them to reflect on these, drawing on their common culture, then to form abstract conceptualizations out of their reflection, and then to apply the abstraction back on their experience.
>
> You can think of it as a circle or clock: At 12 noon students have an experience; at a quarter past they are thinking about it; at half past they are doing some conceptual work around their reflections; and at a quarter to they are doing applications based on their conceptual work. In the Algebra Project this movement from experience to abstraction takes the form of a five-step process that introduces students to the idea that many important concepts of elementary algebra may be accessed through ordinary experiences. Each step is designed to help students bridge the transition from real life to mathematical language and operations.
>
> Because of this connection with real life, the transition curriculum is not only experiential; it is also culturally based. The experiences must be meaningful in terms of the daily life and culture of the students. One key pedagogical problem addressed by the curriculum is that of providing an envi-

ronment where students can explore these ideas and effectively move toward their standard expression in school mathematics.

Understanding Their Power as Active Citizens

As students develop deeper understandings of social and ecological problems that we face, they also often recognize the importance of acting on their beliefs. This notion of nurturing what Henry Giroux has called "civic courage" — acting as if we live in a democracy — should be part of all educational settings, including mathematics classrooms.

Rethinking Mathematics spotlights several examples of student activism. These include fifth-grade Milwaukee students writing letters to social studies textbook publishers based on their mathematical analysis of slave-holding presidents and textbooks' failure to address this issue (see "Write the Truth," page 140); New York City students who measured their school space, calculated inequities, and then spoke out against these inequities in public forums (see "'With Math, It's Like You Have More Defense,'" page 81); and students who used math to convince their school administration to stop making so many obtrusive PA announcements (see the activity "Tracking PA Announcements," page 130).

Motivated to Learn Important Math

Engaging students in mathematics within social justice contexts increases students' interest in math and also helps them learn important mathematics. Once they are engaged in a project, like finding the concentration of liquor stores in their neighborhood and comparing it to the concentration of liquor stores in a different community, they recognize the necessity and value of understanding concepts of area, density, and ratio. These topics are often approached abstractly or, at best, in relation to trivial subjects. Social justice math implicitly tells students: These skills help you understand your own lives — and the broader world — more clearly.

Getting Started

Teachers and preservice teachers sometimes ask: How do I get started integrating social justice concepts in my math class? Our best advice is to take a little at a time. One way is to identify a concept/skill that you are teaching as part of your regular curriculum and relate it to a lesson idea in this book (or on the website for this book, www.rethinkingschools.org/math). Teach the lesson or unit and then gauge how successful it was in terms of student motivation, student understanding of the math concepts, and the deepening of the students' knowledge about the particular social issue.

Another way is to get to know your students and their communities well and listen closely to the issues they bring up. Many of our own social justice projects started from conversations with students about their lives or from knowing about issues in their communities. For example, see the activity "Environmental Hazards," page 41, in which high school students investigate contamination in their own neighborhoods.

The media are also potential sources of projects, because current issues both affect students' lives and have mathematical components that teachers can develop into social justice projects. For example, see "Home Buying While Brown or Black," page 47.

Certainly working in a school that has a conceptually strong foundational mathematics curriculum is helpful. Teachers cannot easily do social justice mathematics teaching when using a rote, procedure-oriented mathematics curriculum. Likewise a text-driven, teacher-centered approach does not foster the kind of questioning and reflection that should take place in all classrooms, including those where math is studied.

By saying this, we do not wish to imply that if teachers use a conceptually based curriculum that embraces the standards put forth by the National Council of Teachers of Mathematics (NCTM) — such as *Investigations* in the elementary grades, *Mathematics in Context* or

Connected Mathematics Project for the middle grades, and *Interactive Mathematics Program* in high school — such a curriculum will automatically guide students towards a social justice orientation. In fact, these programs have an unfortunate scarcity of social justice connections. But a strong, conceptually based foundational curriculum can be a great asset to social justice math teaching, because it can encourage students to critique answers, question assumptions, and justify reasoning. These are all important dispositions toward knowledge that teachers can integrate into their social justice pedagogy.

Occasionally, a teacher needs to defend this kind of curriculum to supervisors, colleagues, or parents. One approach is to survey your state's math standards (or the national standards) and to find references to "critical thinking" or "problem solving" and use those to explain your curriculum. Also, the NCTM clearly states that "mathematical connections" between curriculum and students' lives are important.

But it's important for teachers to recognize that social justice math is not something to sneak into the cracks of the curriculum. It's not something about which we should feel defensive. What we're talking about here is something that helps students learn rich mathematics, motivates them, and is really what math is all about. A social justice approach to math is the appropriate type of math for these unjust times. Other, traditional forms of math are often too abstract, promote student failure and self-doubt, and, frankly, are immoral in a world as unjust as ours. Traditional math is bad for students and bad for society.

Views on Math and Social Justice

The two of us have been teaching math for a combined total of more than 40 years — one of us in a bilingual fifth-grade classroom in a public elementary school and the other in inner-city public middle schools, in alternative high schools, and at the college level. Our perspectives on teaching math for social justice have been shaped by our own involvement in movements for social justice during the past three decades — the Civil Rights Movement, anti-war movements, educational justice movements, and other campaigns. We've also been influenced by educators such as the late Brazilian educator

Math has the power to help us understand and potentially change the world.

Paulo Freire, who argued against a "banking approach" to education in which "knowledge" is deposited into the heads of students and in favor of "problem-posing" approaches in which students and teachers together attempt to understand and eventually change their communities and the broader world.

In addition to the benefits outlined at the beginning of this introduction, an important aspect of a social justice approach to teaching math is that it must include opening up the "gates" that have historically kept students of color, women, working-class and low-income students, and students with perceived disabilities out of advanced mathematics tracks and course offerings. The Algebra Project mentioned above, for example, seeks to ensure that "gatekeeper" classes like algebra don't prevent large numbers of historically disenfranchised students from succeeding in higher education. (For information on other such projects, see Resources, page 163.)

Those who wrote for this book, and those who write for the magazine *Rethinking Schools,* are always encouraged not only to explain what they teach and why they try certain things, but to reflect on how they would do things differently next time. In that spirit we recognize that, as white male educators, our experiences have their own limitations and, if we were to do this book over, we would work harder to increase the representation of authors of color. We encourage all educators who teach math, particularly educators of color, to write about their experiences teaching math for social justice and to consider submitting articles for possible publication in *Rethinking Schools* or online at www. rethinkingschools.org/math, where we share additional teaching ideas, units, and perspectives on teaching math for social justice.

Isn't Math Teaching Neutral?

While reading these articles, some people might question whether it's appropriate to interject social or political issues into mathematics. Shouldn't math teachers and curriculum, they might say, remain "neutral?"

Simply put, teaching math in a neutral manner is not possible. No math teaching — no teaching of any kind, for that matter — is actually "neutral," although some teachers may be unaware of this. As historian Howard Zinn once wrote: "In a world where justice is maldistributed, there is no such thing as a neutral or representative recapitulation of the facts."

For example: Let's say two teachers use word problems to teach double-digit multiplication and problem-solving skills. They each present a problem to their students. The first teacher presents this one:

> A group of youth aged 14, 15, and 16 go to the store. Candy bars are on sale for 43¢ each. They buy a total of 12 candy bars. How much do they spend, not including tax?

The second teacher, meanwhile, offers a very different problem:

> Factory workers aged 14, 15, and 16 in Honduras make McKids children's clothing for Wal-Mart. Each worker earns 43 cents an hour and works a 14-hour shift each day. How much does each worker make in one day, excluding any fees deducted by employers?

While both problems are valid examples of applying multi-digit multiplication, each has more to say as well. The first example has a subtext of consumerism and unhealthy eating habits; the second has an explicit text of global awareness and empathy. Both are political, in that each highlights important social relations.

When teachers fail to include math problems that help students confront important global issues, or when they don't bring out the underlying implications of problems like the first example here, these are political choices, whether the teachers recognize them as such or not. These choices teach students three things:

1. They suggest that politics are not relevant to everyday situations.
2. They cast mathematics as having no role in understanding social injustice and power imbalances.
3. They provide students with no experience using math to make sense of, and try to change, unjust situations.

These all contribute to disempowering students and are objectively political acts, though not necessarily conscious ones.

As high school teacher Larry Steele details in his article "Sweatshop Accounting" (page 53), the seemingly neutral high school accounting curriculum in fact approaches the world in terms of markets and profit-making opportunities. Not everything that counts gets counted, Steele says, and thus the "neutral" curriculum actually reinforces the status quo.

We believe it's time to start counting that which counts. To paraphrase Freida, the ninth grader quoted above, we need to encourage students to defend their rights and to recognize the injustices around them. By counting, analyzing, and acting, we will help students and ourselves better read the world and remake it into a more just place.

Eric (Rico) Gutstein and Bob Peterson
April 2005

PART 1

Viewing Math Broadly

CHAPTER ONE

Teaching Math Across the Curriculum

BY BOB PETERSON

I once read a proposal for an innovative school and it set me thinking about math. It wasn't the proposal's numbers that got my mind going, but rather the approach to structuring math into the curriculum. I disagreed with it.

The plan called for the curriculum to be divided into three areas — math/science, the arts (including fine arts and language arts), and history/philosophy. Blocks of time were set aside for a unified approach in each area.

As I mulled over the proposal and thought of my experience in a self-contained fifth-grade classroom, I realized I was uneasy with the proposed curricular divisions, specifically the assumption that science and math belong together as a unified block. It reminded me of how some elementary teachers integrate the curriculum by lumping language arts and social studies together in one strand, and math and science in another.

It also raised several questions for me. Why place math and science together and not math and social studies? What are the political and pedagogical assumptions behind such an approach? Why shouldn't reformers advocate math in all subject areas? Why not have "math across the curriculum," comparable to "writing across the curriculum"?

One reason reformers have advocated changes in how math is structured is because of the historic problems with math instruction itself: rote calculations, drill and practice ad nauseum, endless reams of worksheets, and a fetish for "the right answer." These have contributed to "number numbness" among students and, ultimately, among the general population when students become adults.

But the problem is deeper than a sterile teacher-centered and text-driven approach. "Number numbness" also has its roots in how math is segregated in schools and kept separate from the issues that confront students in their daily lives. Most students don't want to do abstract exercises with numbers or plod through text-based story problems that have them forever making change in some make-believe store. The curriculum rarely encourages students to link math and history, math and politics, math and literature — math and people. There are unfortunate consequences when math is isolated. First, the not-so-subtle message is that math is basically irrelevant except for achieving success in future math classes, becoming a scientist or mathematician, or making commercial transactions. Second, students learn that math is not connected to social reality in any substantive way. Thus students approach math in the abstract and never are encouraged to seriously consider the social and ethical consequences of how math is sometimes used in society.

Third, if students are not taught how math can be applied in their lives, they are robbed of an important tool to help them fully participate in society. An understanding of math and how numbers and statistics can be interpreted is essential to effectively enter most debates on public issues such as welfare, unemployment, and the federal budget. For example, even though the minimum wage is higher than it's ever been, in constant dollars it is the lowest in 40 years. But you need math to understand that.

When I first began teaching more than 25 years ago, I was dissatisfied with "number numbness," but wasn't sure what to do about it. My growth as a teacher first came in the area of language arts and reading. I increasingly stressed that students should write for meaningful purposes and read books and stories that were connected to their lives. Thus I had children reading and discussing whole books, I conducted writing workshops, and I incorporated reading and writing in science and social studies.

My math, however, remained noticeably segregated from the rest of the curriculum, even though I increasingly emphasized problem-solving and the use of manipulatives.

Later, with the help of my teaching colleague Celín Arce and publications from the National Council of Teachers of Mathematics (see Resources, page 165), I began to view math as akin to language. I now believe that math, like language, is both a discipline unto itself and a tool to understand and interact with the world and other academic disciplines. Just as written and oral language help children understand their community, so can written and oral mathematics. Just as teachers stress the need for "writing across the curriculum," I believe it is important to advocate "math across the curriculum." Just as students are expected to write for meaningful purposes, they should do math for meaningful purposes.

Plans to integrate math into science are a step in the right direction. Assuming that the science curriculum is "meaningful," the teaching of mathematics will improve. But linking math with science is only a beginning and should be followed with integration of math across the curriculum. I have found that my fifth graders, for instance, are particularly interested in social issues. Thus integrating math with social studies is an effective way to bring math alive for the students.

Before I go any further, I want to make two important clarifications. First, I don't mean to imply that distinct math "mini-lessons" aren't important. They are, just as such lessons are necessary in reading and writing. I also want to make clear that integrating math with social studies does not necessarily make the teaching more student-centered or the content more concerned with issues of social justice. Those important components depend on the teacher's philosophical and pedagogical beliefs.

In recent years I have tried in a variety of ways to integrate mathematics — from the simplest understanding of number concepts to solving more complex problems — with social studies. In the interests of clarity (my classroom life is never so neatly ordered), I outline these approaches as: Connecting Math to Students' Lives, Linking Math and Issues of Equality, Using Math to Uncover Stereotypes, and Using Math to Understand History.

Connecting Math to Students' Lives

The starting point for many teachers is to build on what students bring into the classroom, and to connect curriculum to students' lives. Math is a great way to do this. I usually start the year by having kids explore, in small groups, how math is used in their homes and communities. They scour newspapers for numbers, cut them out, glue them onto poster paper, and try to give sense to their meanings, which at times is difficult. They interview family members about how they use math and then write up their discoveries. As part of a beginning-of-the-year autobiography, they write an essay, "Numeric Me," tying in all the numbers that connect to their lives: from height and

weight; to the number of brothers and sisters they have; to addresses, phone numbers, and so forth. I also ask them to write a history of their experiences in math classes, what they think about math, and why.

This process starts a yearlong conversation on what we mean by mathematics and why it's important in our lives. As the class increasingly becomes sensitive to the use of numbers and math in news articles, literature, and everyday events, our discussions help them realize that

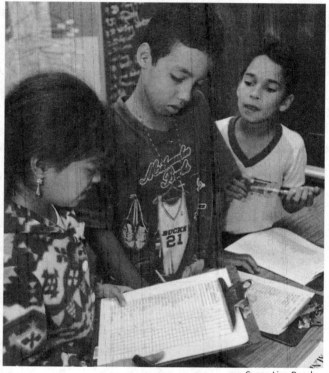

Susan Lina Ruggles

FIFTH-GRADE STUDENTS SURVEY CLASSMATES TO LEARN ABOUT EACH OTHER AND THE POWER OF MATHEMATICS.

math is more than computation and definitions. It includes a range of concepts and topics — from geometry and measurements to ratios, percentages, and probability.

As part of the autobiography project we also do a timeline. We start by putting the students' birthdates and those of their parents and grandparents on a class timeline that circles the outer perimeter of my classroom (and which is used throughout the year to integrate dates that we come across in all subject areas). The students also make their own timelines — first of a typical day and then of their lives. In these activities, students use reasoning skills to figure out relations between numbers, distance, time, fractions, and decimals.

I also use another beginning-of-the-year activity that not only builds math skills but fosters community and friendship. The whole class discusses what a survey or poll is, and then brainstorms questions that they would like to ask each other. First I model how to do a survey, and as a class we graph and write about the data. Then each student surveys her or his classmates about a different topic. My students have, for example, surveyed classmates on their national origins; their favorite fast-food restaurants, music groups, or football teams; and what they think of our school's peer mediation program. Each student tabulates her or his survey data, makes a bar graph displaying the results, and reflects in writing on what she or he has learned. Later in the year students convert the data into fractions and percentages and make circle graphs. I encourage them to draw conclusions from their data and hypothesize about why the results are the way they are. They then present these conclusions orally and in writing.

This activity is particularly popular with my students, and often they want to do more extensive surveys with broader groups of people. The activity thus lays the basis for more in-depth study of polling and statistics around issues such as sampling, randomness, bias, and error. (For extensive curricular ideas on the use of polls and statistics in social studies, refer to *The Power of Numbers* curriculum published by Educators for Social Responsibility. See Resources, page 171.)

Linking Math and Issues of Equality

To help my students understand that mathematics is a powerful and useful tool, I flood my classroom with examples of how math is used in major controversies in their community and in society at large. I also integrate math with social studies lessons to show how math can help us better understand the nature of social inequality. Kids are inherently interested in what is "fair," and using math to explore what is and isn't fair is a great way to interest them in all types of math concepts, from computation to fractions, percentages, ratios, averages, and graphing.

For example, during October and November, there is often lots of discussion of poverty and hunger in my classroom, related either to the UNICEF activities around

POLYP (c) 2005 All rights reserved.

Halloween or issues raised by the Thanksgiving holiday. This is a good time to use simulation exercises to help children understand the disparity of wealth in the United States and around the world. In one lesson (explained in detail in "Poverty and World Wealth," page 64), I provide information on the distribution of population and wealth on the six continents and then have children represent that information using different sets of colored chips.

After I work with students so that they understand the data, we do a class simulation using a map of the world painted on our school playground. Instead of chips to represent data we use the children themselves. I tell them to divide themselves around the playground map in order to represent the world's population distribution. I then hand out chocolate chip cookies to represent the distribution of wealth. As you can imagine, some kids get far more cookies than others and lively discussions ensue. Afterwards, we discuss the simulation and write about the activity.

Not only does such a lesson connect math to human beings and social reality, it does so in a way that goes beyond paper and pencil exercises; it truly brings math to life. I could simply tell my students about the world's unequal distribution of wealth, but that wouldn't have the same emotional impact as seeing classmates in the North American and European sections of the map get so many more cookies even though they have so many fewer people.

I also use resources, such as news articles on various social issues, to help students analyze inequality. In small groups, students examine data such as unemployment or job trends, convert the data into percentages, make comparisons, draw conclusions, and make graphs. This is a great way to help students understand the power of percentages. Because they also use a computerized graph-making program, they real-

ize how the computer can be a valuable tool.

One group, for example, looked at news stories summarizing a university report on 10,000 new jobs created in downtown Milwaukee due to commercial development. According to the report, African Americans held fewer than 8 percent of the new jobs, even though they lived in close proximity to downtown and accounted for 30 percent of the city's population. In terms of the higher-paying managerial jobs, Latinos and

DISPARITIES IN WEALTH

CARTOON TEACHING SUGGESTIONS

Ask students to describe the multiple messages contained in the cartoon below at left. Have them research the distribution of the world's wealth and which companies control major media outlets. Ask why major media corporations might not highlight disparities in wealth. Have students create graphs and cartoons to demonstrate inequality in wealth and who controls major media outlets.

African Americans combined held only 1 percent, while white residents, who are overwhelmingly from the suburbs, took almost 80 percent of the new managerial jobs. Using these data, students made bar and pie graphs of the racial breakdown of people in different jobs and in the city population. They compared the graphs and drew conclusions.

They then did a role play, with some students pretending to be representatives of community organizations trying to convince the mayor and major corporations to change their hiring practices. What began as a math lesson quickly turned into a heated discussion of social policy. At one point, for example, a student argued that one third of the new jobs should be given to African Americans, one third to Latinos and one third to whites, because those are the

ACTIVITY BOX

NO-TV WEEK MATH

During our school's annual "No-TV Week" my students survey the students in our school and use percentages and pie graphs to show, among other things, how many kids have TVs in their bedrooms, how many have cable, who can watch TV whenever they want, and who would rather spend time with their parents than watching TV.

three principal ethnic groups in Milwaukee. Others, however, disagreed. Needless to say, this led to an extensive discussion of what is "fair," why minorities held so few of the jobs created downtown, and what it would take for things to be different.

Using Math to Uncover Stereotypes

It's important for students to be aware of whose voices they hear as they read a history book or the newspaper or watch a movie. Who gets to narrate history matters greatly, because it fundamentally shapes the readers' or viewers' perspective. We can analyze these things with kids and help them become more critical readers of books and other media. In this process math can play an important role.

I usually start with something fairly easy: I have students analyze children's books on Christopher Columbus, tabulating whose views are represented. For instance, how many times do Columbus and his men present their perspective, and how many times do the Taíno Indians present theirs? The students, using fractions and percentages, make large graphs to demonstrate their findings and draw potential conclusions. Large visual displays — bar graphs made with sticky tape, for instance — are good points of reference for discussion and analysis. Math concepts such as percentages, proportions, and comparisons can be used to help kids dis-

cuss the statistics they've uncovered and the graphs they've made.

A similar tabulation and use of percentages can be used to analyze popular TV shows for the number of "put-downs" versus "put-ups," who is quoted or pictured in newspapers, stereotypes of females in popular cartoons, who is included in textbooks, and who is represented in the biography section of the school library. (See "Libraries, Books, and Bias," page 149. Also see "Math and Media," in *Rethinking Our Classrooms, Vol. 1;* see Resources, page 170.)

Using Math to Understand History

As we study history, we pay particular attention to dates and data. I try to highlight numbers that relate to social movements for equity and justice. For example, as we look at the women's struggle for equality we try to imagine what it was like for Susan B. Anthony to go to work as a teacher and get paid $2.50 a week, exactly half the salary of the male teacher she replaced. Lots can be done with such a statistic: for example, figuring out and graphing the difference on an annual or lifetime basis, or looking for wage differentials in other occupations and time periods, including the present. I have found children particularly interested in looking at wages paid to child workers, whether in coal mines or textile mills. We compare such wages to the prices of commodities at the time, to the wages of adult workers, and to the wealth that was accumulated by the owners of industry. Such historical connections can easily be linked to present-day concerns over U.S. child-labor and minimum-wage laws or to international concerns over multinational corporations exploiting child labor in Asia or Latin America to make consumer goods for worldwide markets. An in-depth description of how I have my class approach some of these global issues mathematically and socially can be found in *Rethinking Globalization: Teaching for Justice in an Unjust World* (see Resources, page 170).

One math/history connection, which can range in sophistication depending on the level of the students, is to look at who is represented in different occupations and areas of power in our society, and how that has changed over time. For example, students can figure out what percentage of the signers of the Constitution were slaveholders, common working people, women, wealthy individuals who held bonds, and so forth. A similar exercise would be to analyze U.S. presidents or the people our country has chosen to honor by putting their faces on currency and coins. Such historical number-crunching can take a contemporary turn if the students analyze, for example, the gender and racial breakdown of the U.S. House and Senate, the editors of major newspapers, or the CEOs of Fortune 500 corporations.

It's important for students to understand that such numbers are not permanent fixtures of our social structure but have changed over time as a result of social movements, such as the Civil Rights or women's movements. To demonstrate this, a teacher might have students tally the current percentage of African Americans or women in selected professional occupations and compare it to the 1960s, before the rise of affirmative action.

Another area to teach is the history of math, pointing out the contributions of various non-European cultures and civilizations to mathematical thought. Greek mathematicians, for instance, were heavily influenced by their predecessors and counterparts in Africa and Asia. Arab mathematicians inspired European Renaissance schol-

Susan Lina Ruggles

A STUDENT TOTALS THE RESULTS OF A SURVEY HE CONDUCTED OF HIS CLASSMATES.

ars. The Mayans were one of the first peoples to develop the concept of zero and make sophisticated mathematical calculations. I have used a unit on the base 20 Mayan counting system with my fifth graders to demonstrate such sophistication and to help students expand their understanding of place value. (See the related article "Chicanos Have Math in their Blood," page 70.)

Conclusion

The sophistication and complexity of the math we use in our classrooms naturally depends on the developmental level of our students. Teachers, however, too often underestimate what students are capable of doing. To the degree that I provide quality instruction, clear modeling, and purposeful activities, I am usually pleased with the enthusiasm with which my kids take on such math-based projects and with the success they have in doing them.

I have found that as a result of trying to implement "math across the curriculum" — and in particular, integrating math and social studies — my students' interest and skill in math have increased, both in terms of their understanding of basic concepts and their ability to solve problems. Furthermore, they can better clarify social issues, understand the structures of society, and offer options for better social policies.

Kids need every tool they can get to make this world — their world — a better place. Mathematics is one very important tool. □

DRIVING WHILE BLACK OR BROWN

A MATHEMATICS PROJECT ABOUT RACIAL PROFILING

BY ERIC GUTSTEIN

I did this project in a seventh-grade class, using Chicago and Illinois data, but you can modify it for your setting. The project takes three to five class periods. Start with a discussion of racial profiling and students' experiences with and knowledge of it. End with a discussion of what we can and cannot know from our mathematical analysis, what are our "open" questions, and what are the complexities in understanding racism.

The purpose of this project is to investigate racial profiling, also known as "Driving While Black" or "Driving While Brown." African Americans and Latinos have complained, filed suit, and organized against what they believe are racist police practices: being stopped, searched, harassed, and sometimes arrested solely because they "fit" a racial profile.

But is this true? How can we know? Can mathematics be a useful tool in helping us answer this question?

PART 1: REVIEW BASIC PROBABILITY IDEAS

To understand racial profiling, students need to understand several concepts: randomness, experimentation, simulation, sample size, experimental and theoretical probability, and the law of large numbers (which says the more experiments you run, the closer you come to theoretical probabilities). Acquainting students with these concepts can be as simple as having pairs of students toss a coin 100 times and record their results, and then combining the data from all student pairs and examining how the combined data comes closer to a 50-50 split than do the individual pairs.

PART 2: FIND CHICAGO'S RACIAL BREAKDOWN

Give each group of students a small bag with colored cubes to represent the racial breakdown. I used nine black cubes (for African Americans), nine tan (whites), six red (Latinos), and one yellow (Asians/Native Americans) to approximate Chicago's racial proportions. Do not tell students the total number of cubes or how many of each color there are.

Have students pick one cube from the bag without looking, record its color on the chart (tally marks work well), and replace the cube. Repeat this 100 times. After every 10 picks, have them record the totals on a chart (see page 18). Each line in the chart is the cumulative total of picks. Tell students that they are conducting an experiment (picking/replacing 100 times), collecting data (recording each pick), and analyzing data (determining from their simulation how many cubes of each color are in the bag, the total number of cubes, and the Chicago racial/ethnic percentages).

Make sure students record the fraction and percentage of each race/ethnicity for every 10 picks on the chart.

QUESTIONS FOR EACH GROUP
Emphasize that you want thorough explanations for all questions.

1.   What do you think is in the bag? Why do you think this?

2.   What happened as you picked more times, and what do you think would happen if you picked 1,000 times?

PART 3: INVESTIGATING DWB
Here are sample Illinois data based on police reports from 1987–1997:

•   In an area of about one million motorists, approximately 28,000 were Latinos.

•   Over this period, state police made 14,750 discretionary traffic stops (for example, if a driver changes lanes without signaling or drives one to five miles per hour over the speed limit, police may stop her or him but do not have to).

•   Of these stops, 3,100 were of Latino drivers.

Have students use what they learned in part 2 and set up their own simulation of this situation using cubes (in this example, one could use three different-colored cubes out of 100, or one out of 28, to approximate the ratio of Latino drivers). Have them pick and replace a cube 100 times, record the data, and calculate the results of simulating 100 "discretionary" stops.

MORE GROUP QUESTIONS
3.   What percentage of the motorists in part 3 were Latinos?

4.   What percentage of the discretionary traffic stops involved Latinos?

5.   How did you set up the simulation for part 3 (how many "Latino" cubes and how many total)? Why did you choose those numbers?

6.   In your simulation, how many Latinos were picked out of 100 picks, and what percentage is that?

7.   Do the results from your simulation experiment support the claim of racial profiling? Why or why not?

Combine individual groups' results and analyze as a whole class.

INDIVIDUAL WRITE-UP
1.   What did you learn from this activity?

2.   How did mathematics help you do this?

3.   Do you think racial profiling is a problem, and if so, what do you think should be done about it?

4.   What questions does this project raise in your mind?

End with whole-class discussion. ∎

DRIVING WHILE BLACK OR BROWN

DETERMINING THE RACIAL BREAKDOWN OF THE POPULATION OF CHICAGO

TOTAL # OF PICKS	WHITES			AFRICAN AMERICANS			LATINOS			ASIANS		
	#	fraction	%	#	fraction	%	#	fraction	%	#	fraction	%
10												
20												
30												
40												
50												
60												
70												
80												
90												
100												

CHAPTER TWO

Reading the World with Math

Goals for a Criticalmathematical Literacy Curriculum

BY MARILYN FRANKENSTEIN

When my students examine data and questions, such as the ones on unemployment in the box on page 23, they are introduced to the four goals of criticalmathematical literacy:

1. Understanding the mathematics.
2. Understanding the mathematics of political knowledge.
3. Understanding the politics of mathematical knowledge.
4. Understanding the politics of knowledge.

Clearly, calculating the various percentages for the unemployment rate requires goal number 1, an understanding of mathematics. Criticalmathematical literacy goes beyond this, to also include the other three goals mentioned above.

The mathematics of political knowledge is illustrated here by reflecting on how the unemployment data deepen our understanding of the situation of working people in the United States. The politics of mathematical knowledge involves the choice of who counts as unemployed. In class, I emphasize that once we decide which categories make up the numerator (number of unemployed) and the denominator (total labor force), changing that fraction to a decimal fraction and then to a percent does not involve political struggle — it involves understanding the mathematics. But making the decision of who counts where does involve political struggle — so the unemployment rate is not a neutral description of the situation of working people in the United States. And this discussion generalizes to a consideration of the politics of all knowledge.

In this article, I develop the meaning of these goals, focusing on their interconnected complexity. Underlying these ideas is my belief that the development of self-confidence is a prerequisite for all learning, and that self-confidence develops from grappling with complex material and from understanding the politics of knowledge.

Goal 1: Understanding the Mathematics

I teach at the College of Public and Community Service at the University of Massachusetts-Boston. My students are primarily working-class adults who did not receive adequate mathematics instruction when they were in high school. Almost all my students know how to do basic addition, subtraction, multiplication, and division, although many would have trouble multiplying decimal fractions, adding fractions or doing long division. All can pronounce the words, but many have trouble succinctly expressing the main idea of a reading. Almost all have trouble with basic math word problems. Most have internalized negative self-images about their knowledge and ability in mathematics. In my beginning lessons I have students read excerpts where the main idea is supported by numerical details and where the

THE BOONDOCKS (c) 2005 Aaron McGruder. Dist. by UNIVERSAL PRESS SYNDICATE. Reprinted with permission. All rights reserved.

THE WAR IN IRAQ

CARTOON TEACHING SUGGESTIONS

Have students calculate the cost of health insurance per family based on the data in this cartoon. Have them research the costs of other consumer items or services and create cartoons to compare those costs to that of the war in Iraq. (See pages 29 and 80 for related activities.)

politics of mathematical knowledge is brought to the fore. Then the curriculum moves on to the development of the Hindu-Arabic place-value numeral system, the meaning of numbers, and the meaning of the operations.

I start lessons with a graph, chart, or short reading which requires knowledge of the math "skill of the day." When the discussion runs into a question about a math skill, I stop and teach that skill. This is a non-linear way of learning basic numeracy, because questions often arise that involve future math topics. I handle this by previewing: The scheduled topic is formally taught, while other topics are also discussed. In this way students' immediate questions about other topics are answered, and when the formal time comes for those topics in the syllabus, students will already have some familiarity with them. For example, if we are studying the meaning of fractions and find that in 1985, two hundredths of the U.S. Senate was female, we usually preview how to change this fraction to a percent. We also discuss how no learning is linear, and how all of us are continually reviewing, re-creating, and previewing in the ongoing process of making meaning.

Further, there are other aspects about learning which greatly strengthen students' understandings of mathematics:

- Breaking down the dichotomy between learning and teaching mathematics.
- Considering the interactions of culture and the development of mathematical knowledge.
- Studying even the simplest of mathematical topics through deep and complicated questions.

These are explained in more detail below.

Breaking Down the Dichotomy Between Learning and Teaching Mathematics. When students teach, rather than explain, they learn more mathematics, and they also learn about teaching. They are then empowered to learn even more mathematics. As humanistic, politically concerned educators we often talk about what we learn from our students when we teach. Educator Peggy McIntosh goes so far as to define teaching as "the development of self through the development of others."

Certainly when we teach, we learn about learning. I also introduce research on math education, so that students can analyze for themselves why they did not previously learn mathematics. I argue that learning develops through teaching and through reflecting on teaching and learning. So students' mathematical understandings are deepened when they learn about mathematics teaching as they learn mathematics. Underlying this argument is Paulo Freire's concept that learning and teaching are part of the same process, and are different moments in the cycle of gaining existing knowledge, re-creating that knowledge, and producing new knowledge

Students gain greater control over mathematics problem-solving when, in addition to evaluating their own work, they can create their own problems. When students can understand which questions it makes sense to ask from given numerical information, and can identify decisions that are involved in creating different kinds of problems, they can more easily solve problems others create. Further, criticalmathematical literacy involves both interpreting and critically analyzing other people's use of numbers in arguments. To do the latter you need practice in determining what kinds of questions can be asked and answered from the available numerical data, and what kinds of situations can be clarified through numerical data.

Freire's concept of problem-posing education emphasizes that problems with neat, pared-down data and clear-cut solutions give a false picture of how mathematics can help us "read the world." Real life is messy, with many problems intersecting and interacting. Real life poses problems with solutions that require dialogue and collective action. Traditional problem-solving curricula isolate and simplify particular aspects of reality in order to give students practice in techniques. Freirian problem-posing, on the other hand, is intended to reveal the inter-connections and complexities of real-life situations, even if the specific problems are not solved.

A classroom application of this idea is to have students create their own reviews and tests. In this way they learn to grapple with mathematics pedagogy issues such as: What are the

Real life poses problems with solutions that require dialogue and collective action.

key concepts and topics to include in a review of a particular curriculum unit? What are clear, fair, and challenging questions to ask in order to evaluate understanding of those concepts and topics?

Considering the Interactions of Culture and the Development of Mathematical Knowledge. When we are learning the algorithm for comparing the size of numbers, I ask students to think about how culture interacts with mathematical knowledge in the following situation: Steve Lerman was working with two 5-year-olds in a London classroom. As he recounted:

> [They] were happy to compare two objects put in front of them and tell me why they had chosen the one they had [as bigger]. However, when I allocated the multilinks to them (the girl had eight and the boy had five) to make a tower ... and I asked them who had the taller one, the girl answered correctly but the boy insisted that he did. Up to this point the boy had been putting the objects together and comparing them. He would not do so on this occasion, and when I asked him how we could find out whose tower was the taller he became very angry. I asked him why he thought that his tower was

taller and he just replied "Because IT IS!" He would go no further than this and seemed to be almost on the verge of tears.

At first students try to explain the boy's answer by hypothesizing that each of the girl's links was smaller than each of the boy's, or that she built a wider, shorter tower. But after reading the information, they see that this could not be the case, since the girl's answer was correct. We speculate about how the culture of sexism — that boys always do better or have more than girls — blocked the knowledge of comparing sizes that the boy clearly understood in a different situation.

Studying Mathematical Topics Through Deep and Complicated Questions. Most educational materials and learning environments in the United States, especially those labeled as "developmental" or "remedial," consist of very superficial, easy work. They involve rote or formulaic problem-solving experiences. Students get trained to think about successful learning as getting high marks on school or standardized tests. I believe this is a major reason that what is learned is not retained and not used. Further, making the curriculum more complicated, so that each problem contains a variety of learning experiences, teaches in the non-linear, holistic way in which knowledge is developed in context. This way of teaching leads to a clearer understanding of the subject matter.

Example: In the text below, writers Holly Sklar and Charles Sleicher demonstrate how numbers presented out of context can be very misleading. I ask students to read the text, which is taken from a February 1990 letter published in *Z* magazine, and discuss the calculations Sklar and Sleicher performed to get their estimate of the U.S. expenditure on the 1990 Nicaraguan election ($17.5 million divided by the population of Nicaragua equals $5 per person). This reviews their understanding of the meaning of the operations. Then I ask the students to consider the complexities of understanding the $17.5 million expenditure. This deepens their understanding

ACTIVITY BOX

USING MATH TO TAKE A CRITICAL LOOK AT HOW THE UNEMPLOYMENT RATE IS DETERMINED

BY MARILYN FRANKENSTEIN

In the United States, the unemployment rate is defined as the number of people unemployed divided by the number of people in the labor force. Here are some figures from December 1994. (All numbers in thousands — that is, these figures are one thousandth of the actual total — rounded off to the nearest hundred-thousand).

1. 101,400 employed full time
2. 19,000 employed part time, want part-time work
3. 4,000 employed part time, want full-time work
4. 5,600 not employed, looked for work in last month, not on temporary layoff
5. 1,100 not employed, on temporary layoff
6. 400 not employed, want a job now, looked for work in last year, stopped looking because discouraged about prospects of finding work
7. 1,400 not employed, want a job now, looked for work in last year, stopped looking for other reasons
8. 60,700 not employed, don't want a job now (adults)

In your opinion, which of these groups should be considered unemployed? Why?

Which should be considered part of the labor force? Why?

DO THE MATH

Given your selections, calculate the unemployment rate in 1994.

FOR DISCUSSION

The U.S. official definition counts 4 and 5 as unemployed and 1 through 5 as part of the labor force, giving an unemployment rate of 5.1%. If we count 4 through 8 plus half of 3 as unemployed, the rate would be 9.3%. Further, in 1994 the Bureau of Labor Statistics stopped issuing its U-7 rate, a measure which included categories 2, 3, 6, 7, and 8, so now researchers will not be able to determine "alternative" unemployment rates.

of how different numerical descriptions illuminate or obscure the context of U.S. policy in Nicaragua, and how in real life just comparing the size of the numbers, out of context, obscures understanding.

On the basis of relative population, Holly Sklar has calculated that the $17.5 million U.S. expenditure on the Nicaraguan election is $5 per person and is equivalent to an expenditure of $1.2 billion in the United States. That's one comparison all right, but it may be more relevant to base the comparison on the effect of the expenditure on the economy or on the election, i.e., to account for the difference in per capita income [between U.S. and Nicaraguan citizens], which is at least 30/1 or an equivalent election expenditure in the United States of a staggering $36 billion! Is there any doubt that such an expenditure would decisively affect a U.S. election?

Goal 2: Understanding the Mathematics of Political Knowledge

I argue, as do educators Paulo Freire and Donaldo Macedo, that the underlying context for critical adult education, and criticalmathematical literacy, is "to read the world." To accomplish this goal, students need to learn how mathematics skills and concepts can be used to understand the institutional structures of our society. This happens through:

- Understanding the different kinds of numerical descriptions of the world (such as fractions, percents, graphs) and the meaning of the sizes of numbers.
- Using calculations to follow and verify the logic of someone's argument,

to restate information, and to understand how raw data are collected and transformed into numerical descriptions of the world. The purpose underlying all calculations is to better understand information and arguments, and to be able to question the decisions that were involved in choosing the numbers and the operations.

Example: I ask students to create and solve some mathematics problems using the information in the following article, which appeared in the April 29–May 5, 1992, edition of *In These Times* magazine. Doing the division problems implicit in this article deepens understanding of the economic data and shows how powerfully numerical data reveal the structure of our institutions.

It may be lonely at the top, but it can't be boring — at least not with all that money. Last week the federal government released figures showing that the richest 1 percent of American households was worth more than the bottom 90 percent combined. And while these numbers were widely reported, we found them so shocking that we thought they were worth repeating. So here goes: In 1989 the top 1 percent of Americans (about 934,000

ELECTRICITY COSTS

Declining block-rate payment structure

households) combined for a net worth of $5.7 trillion; the bottom 90 percent (about 84 million households) could only scrape together $4.8 trillion in net worth.

Example: Students practice reading a complicated graph and solving multiplication and division problems in order to understand how particular payment structures transfer money from the poor to the rich.

The "Rate Watcher's Guide," produced by the Environmental Action Foundation, details why under declining block-rate structures low-income citizens who use electricity only for basic necessities pay the highest rates, and large users with luxuries such as trash compactors, heated swimming pools, or central air-conditioning pay the lowest rates. A 1972 study conducted in Michigan, for example, found that residents of a poor urban area in Detroit paid 66 percent more per unit of electricity than did wealthy residents of nearby Bloomfield Hills. Researchers concluded that "approximately $10 million every year leave the city of Detroit to support the quantity discounts of suburban residents." To understand why this happens, use the graph at left below to:

1. Compute the bill of a family which uses 700 kilowatt-hours (kwh) of electricity per month and the bill of a family which uses 1400 kwh.
2. Calculate each family's average cost per kwh.
3. Discuss numerically how the declining block-rate structure functions and what other kinds of payment structures could be instituted. Which would you support and why?

Example: Students are asked to discuss how numbers support Helen Keller's main point and to reflect on why she sometimes uses fractions and other times uses whole numbers. Information about the politics of knowledge is included as a context in which to set her views. Historian Howard Zinn writes:

> Although Helen Keller was blind and deaf, she fought with her spirit and her pen. When she became an active socialist, a newspaper wrote that "her mistakes spring out of the ... limits of her development." This newspaper had treated her as a hero before she was openly socialist. In 1911, Helen Keller wrote to a suffragist in England: "You ask for votes for women. What good can votes do when ten elevenths of the land of Great Britain belongs to 200,000 people and only one eleventh of the land belongs to the other 40 million people? Have your men with their millions of votes freed themselves from this injustice?"

Example: Students are asked to discuss what numerical understandings they need in order to decipher the following data. They see that a recognition of how very small these decimal fractions are, so small that watches cannot even measure the units of time, illuminates the viciousness of time-motion studies in capitalist management strategies.

Economist Harvey Braverman gives samples from time-and-motion studies conducted by General Electric, which were published in a 1960 handbook to provide office managers with standards by which clerical labor should be organized

Opening and Closing Desk Drawers	Minutes
Open side drawer of standard desk	0.014
Open center drawer	0.026
Close side drawer	0.015
Close center drawer	0.027

Chair Activity	Minutes
Get up from chair	0.039
Sit down in chair	0.033
Turn in swivel chair	0.009

Goal 3: Understanding the Politics of Mathematical Knowledge

Perhaps the most dramatic example of the politics involved in seemingly neutral mathematical descriptions is the choice of a map to visualize that world. Any two-dimensional map of

> "In order for us as poor and oppressed people to become a part of a society that is meaningful, the system under which we now exist has to be radically changed. This means that we are going to have to think in radical terms.
>
> I use the term "radical" in its original meaning — getting down to and understanding the root cause. It means facing a system that does not lend itself to your needs and devising means by which you change that system.
>
> That is easier said than done. But one of the things that has to be faced is, in the process of wanting to change that system, how much have we got to do to find out who we are, where we have come from and where we are going.... I am saying, as you must say too, that in order to see where we are going, we not only must remember where we have been, but we must understand where we have been."
>
> — **Ella Baker**, 1969

our three-dimensional Earth will contain mathematical distortions. (See the Mercator and Peters maps, pages 114 and 115.)

The political struggle/choice centers around which of these distortions are acceptable to us, and what other understandings of ours are distorted by these false pictures. For example, the world map with which most people are familiar, the Mercator map, greatly enlarges the size of "Europe" and shrinks the size of Africa. Most people do not realize that the area of what is commonly referred to as "Europe" is smaller than 20 percent of the area of Africa. Created in 1569, the Mercator map highly distorts land areas, but preserves compass direction, making it very helpful to navigators who sailed from Europe in the 16th century.

When used in textbooks and other media, combined with the general (mis)perception that size relates to various measures of so-called "significance," the Mercator map distorts popular perceptions of the relative importance of various areas of the world. For example, when a U.S. university professor asked his students to rank certain countries by size they "rated the Soviet Union larger than the continent of Africa, though in fact it is much smaller," because they were associating "power" with size.

Political struggles to change to the Peters projection, a more accurate map in terms of land area, have been successful with the U.N. Development Program, the World Council of Churches, and some educational institutions. However, anecdotal evidence from many talks I've given around the world suggests that the Mercator is still widely perceived as the way the world really looks.

As Denis Wood writes in his book *The Power of Maps,* "The map is not an innocent witness ... silently recording what would otherwise take place without it, but a committed participant, as often as not driving the very acts of identifying and naming, bounding, and inventorying it pretends to no more than observe."

In a variety of situations, statistical descriptions don't simply or neutrally record what's out there. There are political struggles/choices involved in: which data are collected, which numbers represent the most accurate data, which definitions should guide how the data are counted, which methods should guide how the data are collected, which ways the data should be disaggregated, and which are the most truthful ways to describe the data to the public.

Example: To justify the Eurocentric argument that the Native American population could not have been so great, various "scholars" have concluded that about one million people were living in North America in 1500. Yet, as related in *The State of Native America,* edited by Annette Jaimes, other academics have argued, on the basis of burial mound archeology and other evidence, that the population of the Ohio River Valley alone had been that great, and that "a pre-contact North American Indian population of 15 million is perhaps the best and most accurate working number available." Admitting the latter figure would also require admitting extensive agricultural institutions, as opposed to the less reliable hunting and gathering. Cultivators of land are "primarily sedentary rather than nomadic ... and residents of permanent towns rather than wandering occupants of a barren wilderness."

Example: In 1988, the U.S. Census Bureau introduced an "alternative poverty line," changing the figure for a family of three from $9,453 to $8,580, thereby preventing 3.6 million people whose family income fell between those figures from receiving food stamps, free school meals, and other welfare benefits. At the same time, the Joint Economic Committee of Congress argued that "updating the assessments of household consumption needs ... would almost double the poverty rate, to 24 percent."

Note that the U.S. poverty line is startlingly low. Various assessments suggested that the smallest amount needed by a family of four to purchase basic necessities in 1991 was 155 percent of the official poverty line. John Schwartz and Thomas Volgy wrote in the Feb. 15, 1993, issue of *The Nation* magazine:

> Since the [census] bureau defines the [working poor] out of poverty, the dominant image of the poor that remains is of people who are unemployed or on the welfare rolls. The real poverty line reveals the opposite: A majority of the poor

among able-bodied, non-elderly heads of households normally work full time. The total number of adults who remain poor despite normally working full time is nearly 10 million more than double the number of adults on welfare. Two thirds of them are high school or college educated and half are over 33 [years old]. Poverty in

Statistical descriptions don't simply or neutrally record what's out there.

the United States is a problem of low-wage jobs far more than it is of welfare dependency, lack of education, or work inexperience. Defining families who earn less than 155 percent of the official poverty line as poor would result in about one person in every four being considered poor in the United States.

Example: The U.S. government rarely collects health data broken down by social class. In 1986, when it did this for heart and cerebrovascular disease, it found enormous gaps. As Vicente Navarro wrote in the April 15, 1991, issue of *The Nation:*

> The death rate from heart disease, for example, was 2.3 times higher among unskilled blue-collar operators than among managers and professionals. By contrast, the mortality rate from heart disease in 1986 for blacks was 1.3 times higher than for whites. ... The way in which statistics are kept does not help to make white and black workers aware of the commonality of their predicament.

Goal 4: Understanding the Politics of Knowledge

There are many aspects of the politics of knowledge that are integrated into this curriculum. Some involve reconsidering what counts as mathematical knowledge and re-presenting an accurate picture of the contributions of all the world's peoples to the development of mathematical knowledge. Others involve how mathematical knowledge is learned in schools. British educator Richard Winter, for example, theorizes that the problems so many encounter in understanding mathematics are not due to the discipline's "difficult abstractions," but due to the cultural form in which mathematics is presented. Likewise, Holly Sklar cites a U.S. study that recorded the differential treatment of black and white students in math classes:

> Sixty-six student-teachers were told to teach a math concept to four pupils — two white and two black. All the pupils were of equal, average intelligence. The student-teachers were told that in each set of four, one white and one black student were intellectually gifted and the others were labeled as average. The student teachers were monitored through a one-way mirror to see how they reinforced their students' efforts. The "superior" white pupils received two positive reinforcements for every negative one. The "average" white students received one positive reinforcement for every negative reinforcement. The "average" black students received 1.5 negative reinforcements, while the "superior" black students received one positive response for every 3.5 negative ones.

Class discussion of the above study can lead to examining these ratios and is assisted by creating matrix charts to visualize the data more clearly. Discussion also can involve students in reflecting on topics in mathematics education. This is another example of breaking down the dichotomy between learning and teaching.

Underlying all these issues are more general concerns that should form the foundation of all learning, concerns about what counts as knowledge and why. I think that one of the most significant contributions of Freire to the development of a critical literacy is the idea that "our task is not to teach students to think — they can already think — but to exchange our ways of thinking with each other and look together for better ways of approaching the decodification of an object." This idea is critically important because it implies a fundamentally different set of assumptions about people, pedagogy, and knowledge-creation.

Because some people in the United States, for example, need to learn to write in "standard" English, it does not follow that they cannot express very complex analyses of social, political, economic, ethical, and other issues. And many people with an excellent grasp of reading, writing, and mathematics skills need to learn about the world, about philosophy, about psychology, about justice, and about many other areas in order to deepen their understandings.

In a non-trivial way, we can learn a great deal from intellectual diversity. Most of the burning social, political, economic, and ethical questions of our time remain unanswered. In the United States we live in a society of enormous wealth, yet we have significant hunger and homelessness; although we have engaged in medical and scientific research for scores of years, we are not much closer to changing the prognosis for most cancers. Certainly we can learn from the perspectives and philosophies of people whose knowledge has developed in a variety of intellectual and experiential conditions.

Freire reminds us that currently "the intellectual activity of those without power is always labeled non-intellectual." When we see this as a political situation, as part of our "regime of truth," we can realize that all people have knowledge, all people are continually creating knowledge and doing intellectual work, and all of us have a lot to learn. □

This article was adapted from a version that originally appeared in *Beyond Heroes and Holidays* (see Resources, page 170). For a listing of all citations and references included in the original article, see www.rethinkingschools.org/math.

THE WAR IN IRAQ

HOW MUCH DOES IT COST?

BY BOB PETERSON

Estimating the cost of the war in Iraq is like shooting at a moving target. The costs escalate daily in amounts that many people cannot even imagine. For example, according to the website of the National Priorities Project, www.costofwar.com, by the end of April 2006 the cost of the war was over $277 billion. This is in addition to the nearly $400 billion military budget that the U.S. Congress appropriated for fiscal year 2004–2005.

There is a great deal of debate about how much should be spent on the military so that the United States is well-defended. Some people argue that if money spent on the military was instead spent on combating U.S. and global social and economic problems, we'd all be much safer.

Dwight D. Eisenhower, Commander of Allied Forces in Europe during World War II and the 34th president of the United States (and a Republican), said on April 16, 1953: "Every gun that is made, every warship launched, every rocket fired, signifies in the final sense a theft from those who hunger and are not fed, those who are cold and are not clothed."

TOP GLOBAL PROBLEMS AND THE MONEY NEEDED TO SOLVE THEM

A nonprofit group called the Borgen Project has estimated what it would cost to end certain global problems. Here are some of their estimates for the annual spending needed, based on a projection of 10 years of spending at these levels.:

• Provide shelter for everyone:	$21 billion
• Remove landmines:	$4 billion
• Eliminate nuclear weapons:	$7 billion
• Provide refugee relief:	$5 billion
• Eliminate illiteracy:	$5 billion
• Provide clean, safe water:	$10 billion
• Provide health care and AIDS control:	$21 billion
• Stop deforestation:	$7 billion
• Pay off the debt of poor nations:	$30 billion

continued on next page

DO THE MATH AND QUESTIONS FOR DISCUSSION

1.   Assume you are able to set a $1 bill down on your desk every second without stopping. First have each student in your group record on paper your estimates of how long it would take to set down $1 million, $1 billion, and $156 billion. After recording your guesses, use a calculator to figure out how long each would take. Compare your estimates to your calculated answer.

2.   Using the www.costofwar.com website, estimate how U.S. taxpayers' money is spent on the war each second, each minute, each hour, each day, and each week.

3.   Look at the estimates of how much it would cost to solve the major global issues as defined by the Borgen Project. How does the total cost compare to the cost of the war in Iraq?

4.   A full scholarship to an out-of-state, four-year, major university like the University of Wisconsin–Madison — including dorm fees, food, books, tuition, and other fees — is worth about $27,000 a year. Calculate how many high school seniors could receive a fully paid four-year college education if the monies spent on the Iraq war — as of March 2005 — had been set aside for college scholarships instead.

5.   A group called the National Priorities Project has estimated the amount of taxes that people in different states and cities have paid to fund the war in Iraq. Go to their website, www.nationalpriorities.org, and find your state and/or city. Using this number, calculate how many additional teachers and nurses could be hired in your community, assuming that each of them made the average national salary for their profession. In 2004 the national average teacher salary was $43,250 and the national average nurse salary was $41,642.

6.   Re-read the above quotation from former President Eisenhower. What do you think about the quotation and how does it relate to our current political situation?

7.   Develop a short presentation for the class explaining what you learned and what you think about these matters. ∎

For links to updated spending estimates, budget numbers, and other resources mentioned in this article, see www.rethinkingschools.org/math.

CHAPTER THREE

Race, Retrenchment, and the Reform of School Mathematics

BY WILLIAM F. TATE

The teaching of mathematics needs to be connected to the lives and experiences of African-American students to enable them to fully take part in our democracy. Traditionally, schools have not provided African-American students opportunities to do so.

More than 60 years ago, historian Carter Woodson described this dilemma:

> And even in the certitude of science or mathematics it has been unfor-
> tunate that the approach to the Negro has been borrowed from a "for-
> eign" method. For example, the teaching of arithmetic in the fifth
> grade in a backward county in Mississippi should mean one thing in
> the Negro school and a decidedly different thing in the white school.
> The Negro children, as a rule, come from the homes of tenants and
> peons who live to migrate annually from plantation to plantation,
> looking for light which they have never seen. The children from the

homes of white planters and merchants live permanently in the midst of calculations, family budget, and the like, which enable them sometimes to learn more by contact than the Negro can acquire in school. Instead of teaching such Negro children less arithmetic, they should be taught much more of it than the white children, for the latter attend a graded school consolidated by free transportation when the Negroes go to one-room rented hovels to be taught without equipment and by incompetent teachers educated scarcely beyond the eighth grade.

One important implication of Woodson's argument is that mathematics instruction that is built on a student's life experience provides two mathematics learning environments: within the school and outside the school. Unfortunately, the disciplines that undergird mathematics education — mathematics and psychology — place great stress on objectivity and neutrality. As a result, school mathematics has been tacitly accepted as a color-blind discipline. Thus very little consideration is given to the cultural appropriateness of mathematics pedagogy.

The Current State of Affairs

In recent years, mathematics textbooks have included pictures of African Americans, and some mathematics textbooks provide information about Africans and African Americans who have contributed to the development of the discipline of mathematics. These efforts represent progress and should be encouraged. Yet I doubt seriously if these efforts will prove sufficient to enfranchise African-American students in mathematics. I contend that connecting the pedagogy of mathematics to the lived realities of African-American students is essential to creating equitable conditions in mathematics education.

What type of pedagogy must African-American students negotiate to be successful in school mathematics? Conventional mathematics pedagogy emphasizes whole-class instruction, with teachers modeling a method of solving a problem and students listening to the explanation. This is typically followed by having the students work alone on a set of problems from a textbook or worksheet. The goal of this teacher-directed model of instruction is for students to produce correct responses to a narrowly prescribed problem. This type of mathematics pedagogy is consistent with several studies of mathematics instruction conducted in the 1970s.

Unfortunately, this conventional mathematics pedagogy is exactly the kind of "foreign method" of teaching described by Woodson. Few if any attempts are made to build on the thinking and experiences of African-American students. One important purpose of mathematics education is to prepare students to incorporate mathematical reasoning and communication into their everyday lives. However, conventional pedagogy has often persuaded students — particularly African-American students — to consider school mathematics as a subject divorced from their everyday experiences and from their attempts to make sense of their world.

Today, the "foreign" pedagogy delineated by Woodson continues in many different forms. For example, pedagogy for African-American students is hampered by the following conditions of their schooling:

- Persistent tracking.
- Less access than other students to the best-qualified teachers of mathematics.
- Fewer opportunities to use technology in school mathematics.

Moreover, as the proportion of African-American students in a school increases, the relative proportion of college-preparatory or advanced sections of mathematics decreases.

What is the effect of the foreign pedagogy

on African-American students' thinking and achievement in mathematics? Data from the National Assessment of Educational Progress (NAEP) indicate that African Americans, across the grade levels tested, scored significantly better on portions of the test related to factual knowledge and basic computational skills than they did in previous years. However, no growth was evident on portions of the test that assessed more advanced levels of mathematical reasoning. Should we be surprised at these findings? Walter Secada commented on the consistency of the NAEP findings with the pedagogical focus of the previous decade:

> Indeed, if we agree that the 1970s and 1980s were times when educational policy was predicated on mastery of basic skills, we could read these data as telling a success story: Insofar as we had set a national goal — the acquisition of basic skills — we moved in the direction of its attainment.

Susan Lina Ruggles

IN CONVENTIONAL MATH PEDAGOGY, LITTLE EFFORT IS MADE TO BUILD ON THE EXPERIENCES OF AFRICAN-AMERICAN STUDENTS.

African Americans did become proficient in their mastery of low-level basic skills. Alter-nately, we might read this as a story of incomplete success: Insofar as basic computation skills are deemed insufficient for the knowledge and mastery of mathematics, these data document how much father we have to go.

Race, Reform, and Retrenchment

What role should the reform of school mathematics play in changing this story of incomplete success? Reform documents pub-

lished by the National Council of Teachers of Mathematics have called for a new vision of mathematics pedagogy. For example, the publication *Professional Standards for Teaching Mathematics* recommends that all mathematics teachers know "how students' linguistic, ethnic, racial, gender, and socioeconomic background influence their learning of mathematics," and "the role of mathematics in society and culture, the contribution of various cultures to the advancement of mathematics, and the relationship of school mathematics to other subjects and realistic application."

Despite such recommendations, mathematics pedagogy built on the lives and experiences of African-American students must overcome many formidable barriers if it is to make its way into our classrooms. Many scholars argue that the curriculum and pedagogy of mathematics have been, and continue to be, connected to Eurocentric precepts that exclude the African-American experience. For example, George Joseph states:

> The progress of Europe and its cultural dependencies [including the United States of America] during the last 400 years is

"Almost all the experiences discussed in American classrooms are approached from the standpoint of white perspectives."

perceived by many as inextricably — and even causally — linked with the rapid growth of science and technology. So in the minds of many, scientific progress becomes a uniquely European phenomenon, to be emulated only by following the European path of social and scientific development.

Joseph's remarks are germane to the current debate involving equity in the reform of school mathematics. Rather than address the problem of how to connect mathematics to the lives of African-American students, many advocates of reform have argued that African-American students warrant better treatment in mathematics classrooms on the grounds of national economic interest and global competition. For example, a report from the Carnegie Commission on Science, Technology, and Government states:

> The national interest is strongly bound up in the ability of Americans to compete tech-

nologically. This requires not only an adequate supply of scientific and technical professionals but a workforce able to solve problems and use the tools of a knowledge-intensive economy. All young people, including the non-college bound, the disadvantaged, and young women must be given the opportunity to be-come competent in science and mathematics.

Arguments based on the need to prepare workers for life in a global economy, while persuasive to the public, generally overlook the problems facing African-American students in many school settings. Michael Apple suggests that these appeals to economic self-interest are attempts to persuade educators to compromise their beliefs about individual human rights because of the current fierce global competition for economic prosperity. The tension between mathematics pedagogy that is connected to the experiences of African-American students (an individual human right) and the national economic interest is real and can be a barrier preventing educators from seeking equity in mathematics education.

The documents that have spawned the movement to reform mathematics education have called for a new era of "mathematics for all." These documents have embraced the idea of improving the mathematics performance of African-American students in our nation's urban centers, but achieving this objective will require a blueprint. As Larry Cuban warned, "Unless policymakers and practitioners begin to consider how problems involving schools are framed, they will continue to lunge for quick solutions without considering the fit between the solution and the problem."

The current reform movement in mathematics education has been framed within a discussion of national economic interests. This focus raises questions about the ability of current reforms to generate interest in pedagogical practices that go beyond those intended to yield gains for the national economy. I contend that the educators and policymakers leading the mathematics reform movement should address two questions related to the problems facing African-American students in mathematics classrooms across the United States. First, should school districts support (and teachers provide) mathematics teaching built on the experiences and lives of African-American students? And second, should the focus of mathematics teaching be to prepare African-American students to participate in the national economy or in our democracy? Both questions are extremely complex and probably do not have one right answer. However, I will devote the remainder of my discussion to examples that help illuminate the urgency of these questions.

"Centricity" and Pedagogy

I have argued that one barrier to an equitable mathematics education for African-American students is the failure to "center" them in the process of knowledge acquisition and to build on their cultural and community experiences. Molefi Asante defines the concept of centricity as "a perspective that involves locating students within the context of their own cultural references so that they can relate socially and psychologically to other cultural perspectives." The concept of centricity can be applied to any culture. "For the white students in America," Asante continues, "this is easy, because almost all the experiences discussed in American classrooms are approached from the standpoint of white perspectives and history."

The idea that mathematics pedagogy should be built on the experiences of the student is undergoing a resurgence in the current reform movement. Yet very little has been written about building a mathematics program centered on the thinking and experiences of

African-American children. I offer two examples of the influence of mathematics pedagogy that has not been constructed to center the African-American child.

The first example is based on an experience with a student teacher at my university. She provided her class with the following problem: Joe has five pumpkin pies. Karen has six pumpkin pies. How many pumpkin pies do Karen and Joe have all together?

This problem was written on the chalkboard when I entered the classroom. Five children were seated in the room — one African American and four whites. This was a second-grade class at the beginning of a school year. These five students were provided separate mathematics instruction so that they could receive individualized attention to help improve their performance. The four white children were busy using manipulatives and appeared excited about the process of solving the problem. On the other hand, the African-American child was not engaged in any outward activity. He was very quiet and appeared uninterested.

I asked the student teacher about the African-American student's behavior. She replied, "Mark does not like math. I don't understand why, he's not a bad kid." Curious, I inquired about the discussion that led to the pumpkin pie math problem. The student teacher responded, "I used pumpkin pie as the object to be added because Thanksgiving is in two weeks." She went on to say that she wanted to motivate the students to solve a problem related to their holiday experience. In essence, she was trying to center the students in the process of knowledge acquisition. She was using a cultural construct — Thanksgiving pumpkin pie — as the means of centering her students. Yet her attempt to center the African-American student was unsuccessful.

I asked the student teacher if she thought every family ate pumpkin pie on Thanksgiving. Her response: "Probably." I asked her to ask each of the five children in the class. She discovered that pumpkin pie was indeed a Thanksgiving ritual in the homes of the white children. For the African-American child, however, sweet potato pie was the dessert of the day. Thus the background discussion that led to the problem was "foreign" to this student.

The student teacher had worked to provide the children in her class with what she thought was a problem centered in the context of their lives. Yet the problem reflected the

> "Grandma understood many things that are barely recognized in the wider educational world even today. For example, she realized that arithmetic is injurious to young minds and so, after I had learned my tables, she taught me algebra. ... She thought that ... drill was stultifying. The result was that I was not well-drilled in geography and spelling. But I learned to observe the world around me."
>
> **— Margaret Mead,**
> reflecting on her own education at home

default position of an idealized white middle-class reality. The mathematics problem and her pedagogy were unwittingly constructed to be the property of white children.

Understand my point here: I am not suggesting that all African-American families celebrate Thanksgiving in the same way or that all white families do either. Rather, I contend that the default position of mathematics curriculum, assessment, and pedagogy is often more closely aligned with the idealized experience of the white middle class. Moreover, this reality is subtle and often difficult to diagnose. The diagnosis is difficult because, traditionally, mathematics has been viewed as neutral and objective.

Implicit Assumptions

This leads to my second example: A group of teachers from an urban middle school discovered this very point when a large number of

students at their predominantly African-American school responded "strangely" to an assessment item on a districtwide mathematics test. The basic structure of the test item was as follows: It costs $1.50 each way to ride the bus between home and work. A weekly pass is $16. Which is the better deal, paying the daily fare or buying the weekly pass?

The district's test designers constructed the problem on the assumption that students who solved the problem correctly would choose to pay the daily fare. Implicit in the design of the test item is the notion that all people work five days a week. It is also assumed that the employee has only one job. Yet these assumptions are not consistent with the daily lives and realities of many African-American students. Thus it should not have been shocking that a large percentage of the students in this particular middle school thought that buying the weekly pass was the better option.

When school officials questioned the students about their responses, they found that many of the students had centered themselves in the solution process. These students converted the "neutral" context of the problem into their own property. For instance, the students commented on the fact that more than one family member could use a weekly pass. They also mentioned the option of using the weekly pass on Saturdays and Sundays. In the families of many African-American students, the financial providers hold several jobs, and work both on weekdays and on weekends. For these students, choosing the weekly pass is economically appropriate and mathematically logical.

Taking a "centric" perspective would enable mathematics teachers to expect their students' responses to be linked to the specific details of the students' lives. Unfortunately,

African-American students who center their lives and experiences within the process of acquiring knowledge about mathematics risk being put down for focusing on "extraneous matters." As a result, many African-American students view mathematics as a subject appropriate only for white males, and they fail to see the relevance and usefulness of the discipline.

A closer look through a centric framework reveals important details about the Eurocentric terrain that the children in this urban middle school had to negotiate in order to claim ownership of the mathematics presented on their test. First, the students had to understand that the mode of response should reflect the idealized experiences of a traditional white middle-class family. Second, the students had to understand the mathematics involved in solving the test item using the white middle class as a frame to guide their problem solving. This dual consciousness was required if the students were to "succeed" according to the Eurocentric assumptions of the test designers.

Dominant Culture

Is the first level of consciousness fair or empowering to African-American students whose experiences are not those of the white middle class? William Harvey provides what I consider an appropriate response to this question.

> The advocacy of the values and lifestyles of the dominant white culture is not and cannot be psychologically beneficial to blacks. The dominant cultural value system, and that of the school, admonishes black students for being what they are, while physical and social reality prevents them from being anything else.

In order to take advantage of the diversity in a single classroom, a mathematics teacher's pedagogy should try to provide students with opportunities to solve problems using their experiences. Of course, the mathematical processes used to solve problems should be consistent with the contexts of students' lives.

Furthermore, mathematics teachers working from a centric perspective would endorse having their students solve the same problem from the perspective of different members of the class, school, and society. This approach is consistent with the methods of successful teachers of African-American children. Gloria Ladson-Billings found that these teachers looked to connect students' knowledge of self with broader social and political realities.

Conflict, Democracy, and Pedagogy

A second barrier to providing an equitable mathematics education for African-American students is the failure to prepare them for the conflicts of democracy.

Traditionally, mathematics education has been connected to issues of national economic survival, rather than to the development of democratic citizenship through critical thinking in mathematics. The latter involves helping students engage in mathematical thinking that is connected to the students' social and political contexts. The intent of this pedagogy is twofold: First, students are being prepared to take an active role in our democracy; and second, students are provided important insight into, and understanding of, the role of mathematics within the democratic system of governance.

Until recently, embedding mathematics pedagogy within social and political contexts was not a serious consideration in mathematics education. The act of counting was viewed as a neutral exercise, unconnected to politics or society. Yet when do we ever count just for the sake of counting? Only in school do we count without a social purpose of some kind. Outside of school, mathematics is used to advance or block a particular agenda. However, mathematics curriculum and pedagogy rarely prepare the African-American student to engage in authentic contexts of democracy.

There is a growing consensus among scholars that knowledge is constructed through the interaction of the mind and authentic experiences. For example, John Seely Brown and his colleagues have theorized that in-school knowledge is acquired by working alone, memorizing static rules for well-defined problem settings. Acquiring knowledge of mathematics in this type of setting makes transfer difficult. In contrast, out-of-school knowledge is obtained by working in social settings in which it is important to know the reason for a conflict, to solve ill-defined problems, and to construct personal (i.e., centric) meaning. The activity, context, and culture of the learning environment are refer-

are negatively affirming their community, including the 13 liquor stores within a 1,000-foot radius of the school. Each member of the class has been greatly influenced by the presence of the liquor stores (i.e., the problem is centric). For example, one female student commented: "We pass by drunks asking us for money. We keep going but they harass us and tell us to come here."

The students have found that the disproportionate number of liquor stores in their community is a function of local legislation. To resolve the conflict between the local liquor laws and their negative experiences, the students have formulated and proposed mathematically based economic incentives to get liquor stores to relocate away from the school.

For Mason's students, mathematics is thus more than just numbers shorn of context (i.e., in-school knowledge). Instead, mathematics is embedded within the culture of their daily struggles. These students are learning to use mathematics to support their political positions. For example, the students in this class have mastered the relationship between media "sound bites"

Susan Lina Ruggles

WORKING TOGETHER IN SOCIAL SETTINGS TO INVESTIGATE REAL-LIFE PROBLEMS, MOTIVATES STUDENTS TO ACQUIRE OUT-OF-SCHOOL KNOWLEDGE.

ence points for the retrieval, interpretation, and use of mathematical knowledge. A child's construction of meaning is a result of social interactions — the negotiation of understanding that derives from engaging in real activities of the culture.

Perhaps an example of the work of a practicing teacher will illustrate how mathematics pedagogy can build on the personal and social perspectives. The "War Room" is occupied by Sandra Mason's students. The 10 students in her class have targeted 25 problems that they feel

and mathematics, because Mason has prepared them to ask and answer several questions, using mathematics, that will be important in launching arguments for their position or countering the arguments of their opponents. For example: What methods of mathematical analysis will best support our position? What variables should be included in our analysis to strengthen our position? How can we minimize the influence of variables that may weaken our position? Will percentages or raw numbers make a more striking impression?

These questions are integral aspects of "mathematizing" in our society. Yet such questions are rarely found in the sources that guide the development of mathematics reform. An African-American child will confront similar "mathematized" renditions of social reality every day. For instance, the battle over the appropriate use of racial categories in the U.S. census is still being waged. The struggle over the construction of racial categories for the census is one of the most technically complex and socially relevant debates of our time. When will African-American and other children encounter such a debate in school mathematics? Will the traditional curriculum and pedagogy of mathematics prepare our children to ask the types of questions that are necessary for participation in a democracy?

Mathematical Production and Conflict

I propose the need for a mathematics curriculum and pedagogy that recognize mathematical production as a function of conflict on two levels.

First, like all communities, the mathematics community is governed by stated rules and a covert value system. Within this community, individuals and groups of mathematicians have a significant history of struggle over what is deemed appropriate knowledge for the discipline. The many other issues over which battles have raged within this community include the interpretation of data, credit for discovering ideas, and censorship. The very nature of mathematical production involves conflict. Yet traditional mathematics classrooms are structured to rank students' understanding of a body of static ideas and procedures. Hidden from the students is the role of conflict in advancing mathematical thought.

Second, the use of mathematics in our democratic society is almost always linked to an attempt by one group or individual to secure control of property. This means a social, political, and/or economic conflict will necessarily ensue.

In economic terms, there are two conditions at work when issues are mathematized.

> The use of math in our democratic society is almost always linked to an attempt to secure control of property.

First, situations are mathematized in order to maximize the return on the information that is being analyzed. Thus mathematics pedagogy should seek to prepare students to answer the following question: What agenda is being maximized by a given example of mathematizing?

For example, hospitals have experimented with computer systems that can calculate the probability of surviving an illness, based on specific variables about the patient. This development provides teachers with an opportunity to explore with their students a mathematized situation. For instance, decisions to place patients on life-support systems could be influenced by the mathematical models used by the computer. Mathematizing the decision to place a patient on life support — under the banner of optimizing system use, minimizing hospital cost, and maximizing profit — could result in discriminatory practices. Adopting the personal (centric) and social (conflict) perspectives would lead mathematics teachers to explore a variety of questions with their classes. What variables are included in this model? How do variables such as gender, family history, and income influence a patient's chances of receiving life-support services? Does being an African-American male lower a patient's chances of receiving life-support services, since, comparatively speaking, African-American males have a shorter life expectancy?

In essence, does the computerized mathematical model reinforce existing survival rates?

The second reason situations are mathematized is to minimize the risk of successful challenges to the decision-making process. How many students leave school with enough knowledge and practice to challenge the use of mathematics in society? Thus far, it has been the case that those few who have learned to use mathematics as a tool to guide their own decision-making have stifled the voices of the large segment of the population that does not know how to counter with their own mathematically based arguments — a disproportionate number of whom are African Americans.

Learning to Analyze Mathematized Situations

If students learn in school to analyze and critique mathematized situations, such as the hospital computer system described above, they will be prepared for public discussions about the development and implementation of the new models that are used in social decision-making. Such preparation is radically different from merely preparing students to add, subtract, multiply, and divide accurately. Our highly technological society requires that all, not just African-American, students be prepared to use mathematics to defend their rights. The curriculum and pedagogy of mathematics should support this objective.

I have argued that African-American students should be provided a mathematics pedagogy that is built on their cultural experiences. Furthermore, I contend that this mathematics pedagogy should prepare African Americans and others to participate in our democracy. Traditionally, these two recommendations have not been part of the "mainstream" discourse on school mathematics. More recently, standards for school mathematics reform have called for pedagogy that recognizes these objectives. However, research suggests that many educators will most likely implement only those mathematics standards that are consistent with their own beliefs about appropriate mathematics pedagogy.

Will teachers and other educators view as important the dual need to build on students' personal experiences and to empower them to function in a democracy? Will they move to change current practice? Woodson provides an appropriate ending to this discussion:

> In the first place, we must bear in mind that the Negro has never been educated. He has merely been informed about things which he has not been permitted to do. The Negroes have been shoved out of the regular schools through the rear door into the obscurity of the backyard and told to imitate others whom they see from afar, or they have been permitted in some places to come into the public schools to see how others educate themselves. The program for the uplift of the Negro in this country must be based upon the scientific study of the Negro from within to develop in him the power to do for himself.

The prospect of a new beginning for mathematics education rests with the ability of mathematics teachers to provide pedagogy that builds and expands on the thinking and experiences of African-American students. Moreover, this pedagogy should focus on preparing these students to function within our democracy. □

This article is adapted from a version that originally appeared in *Phi Delta Kappan* magazine. For a link to all references and citations included in the original article, see www.rethinkingschools.org/math.

ENVIRONMENTAL HAZARDS

IS ENVIRONMENTAL RACISM REAL?

BY LARRY MILLER

This activity makes use of an interactive online map maintained by the U.S. Environmental Protection Agency (EPA), which shows the location of environmental hazards in considerable detail.

The purpose of this activity is to investigate whether environmental racism is real.

PROCEDURE

1. With students, identify the zip codes in the area in which they live (or in some urban area).

2. Go online to the interactive map at www.epa.gov/enviro/wme. Type in a zip code or city and state, and click the "Create My Window" button. A map of the selected area will appear.

3. Click the "regulated sites" button in the left-column list. This will bring up a sub-list including "multiple activities," "Superfund," "toxic release," "water discharge," "air emissions," and "hazardous waste."

4. Click on each of these areas. Then click "redraw map" at the top of the left column. A new map will appear showing all the identified EPA sites in that area.

5. Have students follow the procedure for steps 6 through 10 below.

6. Count the number of city blocks shown on the map from east to west, and divide that number by 12 to determine the length in miles. Do the same from north to south.

7. Multiply the length by the width in miles to get the size of the zip code area in square miles.

8. Count the number of identified EPA sites (all colors) on the map.

9. Put the number of sites in ratio to the square miles. This is the density (number of sites per square mile.)

10. Divide the number of sites in a zip code by the square miles. This will give density per square mile. This ratio is needed to compare zip code areas because of their different sizes.

continued on next page

11.   Repeat steps 2 through 10 using the zip code of an adjacent suburban area.

12.   Compare the density of hazardous sites in students' neighborhoods or the urban area to the density in the suburbs. (In Milwaukee, students see a huge disproportion when they compare the ratios found in the city and the suburbs.)

13.   Have students read this definition of environmental racism:

> Environmental racism is racial and class discrimination in environmental policymaking and the enforcement of regulations and laws; it is the deliberate targeting of people of color and their communities for toxic and hazardous waste facilities and the sanctioning of the life-threatening presence of poisons and pollutants in poor communities and communities of color.

14.   Students then answer and discuss the question: "Is environmental racism real?"

15.   Have students visit the Voices from the Grassroots website, www.ejrc.cau.edu/voicesfromthegrassroots.htm, to read reports from around the country about the environmental justice movement and report on their findings to the rest of the class. ∎

For a detailed description of how one class took action based on this lesson, see www.rethinkingschools.org/math.

STUDENTS FROM LARRY MILLER'S CLASS HELPED EXPOSE A TOXIC MESS IN THEIR NEIGHBORHOOD.

CHAPTER FOUR

Historical, Cultural, and Social Implications of Mathematics

BY S. E. ANDERSON

The Egyptians, Chinese, and Indians used different styles of mathematical generalization in algebraic problem-solving. Pre-Hellenic algebraic proofs, such as those found in two of the most important Egyptian mathematical documents, the Ahmes Papyrus (c. 1650 B.C.) and the Moscow Papyrus (c. 1850 B.C.), while not deductively axiomatic, were and still are valid proofs.

The Egyptians created a tremendous civilization that lasted thousands of years. Egypt is in Africa, and the people who inhabit the land were and are Africans. The Egyptian civilization produced complex technological innovations and forms of communication, and engaged in an extensive interchange of goods and ideas with other people, thousands of years before they helped bring forth Greek civilization. At least half of the Greek language is African in origin and the Greek cosmological and mythological constructs were founded upon Egyptian constructs, as was Greek shipbuilding, architecture, and mathematics.

Euclid — the so-called "father" of plane geometry — spent 21 years studying and translating mathematical tracts in Egypt. Pythagoras also spent years studying philosophy and science in Egypt, and possibly journeyed East to India and/or Persia, where he "discovered" the so-called Pythmatical documents (c. 800–500 B.C.). How could a theorem whose proof was recorded in Babylonian documents dating 1,000 years before he was born be attributed to Pythagoras?

We need to shatter the myth that mathematics was or is a "white man's thing," and to show that all civilizations, though they differ and develop at different paces, have always been bound inextricably to each other. It is important that students know that Europe is not now, nor was it ever, the "civilizing center" of the world surrounded by wildness and chaos.

To further undo this Eurocentric assumption, students need to know about the constant flow of ideas and techniques into Europe from the early Greeks through the Medieval and Renaissance periods to the rise of capitalism. Certain aspects of European mathematics could not have developed had not the Europeans traded with more advanced societies. One of the most glaring examples of this is the case of the 150-year political struggle around the incorporation of the Hindu-Arabic numeral system into common usage in Europe. For a century and a half during the Medieval period (specifically around 1200–1350 A.D.), the dominant Roman Catholic Church's fear of a rising rival class, coupled with European racism and xenophobia, impeded the spread of mathematical knowledge throughout Europe. The Vatican denounced Hindu-Arabic numerals as "the work of the devil" because it viewed the widespread use of an easy way to calculate as a means by which European merchants and craftsmen would become even more independent of the Church.

The example leads into a discussion of the myth of the "Dark Ages," which asserts that because a general retrogression occurred among the European feudal elite during the Medieval period, nothing was happening intellectually anywhere else in the world. On the contrary, ideas in mathematics, science, and philosophy flourished, both inside and outside of Europe during that time. Great African and Middle-Eastern Arab scholars lived and studied in places like Toledo and Cordoba in Spain as well as in Sicily, Timbuktu, Cairo, Baghdad, and Jundishapur, which were also key centers of scholarly learning and research. The rich and complex Arab culture that dominated southern Europe, most of Africa, the Middle East, and parts of India and China during this period brought forth such intellectual centers as Caliph al-Mansur's House of Wisdom (Bait al-Hikma) in Baghdad, where documents, scholars, and researchers integrated the great astronomical studies of Indian, Chinese, Greek, and Babylonian scholars.

Bait al-Hikma was also a research university out of which a key mathematician evolved: Mohammed ibn-Musa al-Khwarizmi (c. A.D. 825–?). He authored two foundational mathematics texts: The first, *Hisab al-djabr wa-al Muqabala* (*The Science of Equations* or *The Science of Reduction and Cancellation*) not only gave Europe its first systematic approach to algebra, but also was the source of the name for the subject matter (al-djabr or "algebra"). His second book (now found only in the Latin original), *Algorithmi de Numero Indorum,* explains the Indian origins of the numeral system. Subsequent European translations of this book attributed the system to him; hence, schemes using these numerals came to be known as "algorithms" (a corruption of the name "al-Khwarizmi"), and the numerals became known as "Arabic numerals." □

This article is adapted from a piece that first appeared in the *Journal of Negro Education* and later in "Worldmath Curriculum: Fighting Eurocentrism in Mathematics" in *Ethnomatics: Challenging Eurocentrism in Mathematics Education,* edited by Arthur B. Powell and Marilyn Frankenstein.

PART 2

Infusing Social Justice into Math Classes

CHAPTER FIVE

"Home Buying While Brown or Black"

Teaching Mathematics for Racial Justice

BY ERIC GUTSTEIN

So one question leads to another question, and then you have to answer four more, and those four questions lead to eight more questions. So I think that [racial disparity in mortgage lending] is not racism, but that leads me to the conclusion that if it was not racism, then why do they pay more money to whites? Is that racism?

— Vanessa, seventh grade

Vanessa wrote these words on a mathematics project I taught to middle school students titled "Mortgage Loans — Is Racism a Factor?" To begin the project, I had the students read a *Chicago Tribune* article that analyzed mortgage rejection rates for African Americans, Latinos, and whites in

68 different metropolitan areas. The students mathematically analyzed data and wrote essays about whether racism was a factor, using data and arguments from the article. The mortgage project is an example of how I use mathematics to teach for racial justice.

A central part of teaching for social justice is to work for a society where racism is reduced and, eventually, eliminated. But racism is so deeply embedded in the history, consciousness, and fabric of U.S. life that to remove it will take long-term, concentrated efforts. We need to understand racism, its genesis and manifestations, and also build schools and classrooms that explicitly promote racial justice (and that are linked to anti-racist social movements on a broader scale).

I have occasionally taught middle school mathematics in a Chicago public school, located in a low-income, Mexican immigrant community, as part of my job as a university-based mathematics educator. I use ideas of teaching for social justice along with helping students develop mathematical power (being able to reason and communicate mathematically, develop their own mathematical thinking, and solve real-world problems in multiple and novel ways) — and pass the "gatekeeping" standardized tests.

Through studying these experiences and other learnings, I have come to certain ideas about teaching mathematics for racial justice. In my view, a racial justice curriculum should provide students opportunities to:

- Develop an understanding of the sociopolitical, cultural, economic, and historical dynamics of racism, along with their interconnections.
- Appreciate the complexity of different forms of racism (structural, institutional, individual).
- Develop and use analytical tools (such as data analysis, graphing, and mathematical modeling of social phenomena) to understand and dissect racism.
- Support views, develop coherent arguments, and engage in group discussions to develop individual and collective analyses.
- Develop an appreciation for multiple, alternative, and competing perspectives while constructing their own independent knowledge.
- Become active, get involved in actual struggles when possible and appropriate, and use their analytical tools to cut through and confront myths and historical inaccuracies.

Another important issue for teachers is that classroom cultures supporting investigations of complex and emotional issues like racism have to be co-created with students over time. One cannot just "plop down" potentially volatile issues without creating conditions for students to take seriously their roles as learners and knowledge creators. Students need to be able to listen to others and find their own voices as well. In highly regulated urban schools, creating classroom norms that support this type of work can be challenging and takes time, because such a culture contradicts students' expectations of what teaching "should be" — and students are not always so quick to accept these new routines and new topics of discussion.

This last point is particularly important when white teachers are teaching about racism and for racial justice with students of color. As a white teacher of students of color, I consider the following questions: Why should students of color, who know firsthand the pain and horror of racism, necessarily share their thoughts, feelings, and ideas in a classroom with a white teacher? What difficulties exist when teaching about racism across color lines, given sociopolitical and cultural power differentials in a racist society?

Since my students have direct experience with racism, and my knowledge is more analytical, historical, and observational, how do we meet in mutual respect to deepen our collective understanding of the impact of racism and how to fight better against it? How do I better understand how my students perceive and experience racism and other forms of discrimination so that I can create opportunities for them to better comprehend the nature of these injustices — all while teaching mathematics in a high-stakes, urban school district focused on "accountability," regimentation, and discipline?

I have found it challenging, but possible, to try to enact the above principles in the classroom. What follows is a lesson in which I tried to put these ideas into practice.

An Example of Racial Justice Curriculum

In December 2002, I gave my students the mortgage project. (Materials for this project are available online at www.rethinkingschools.org/math). This lesson was challenging because the article from the *Chicago Tribune* was confusing, the mathematics was complicated, and terms were used without definition. But it was clear that in Chicago, African Americans were rejected for mortgage loans five times as often as whites, and Latinos were rejected for mortgage loans three times as often as whites. This was true across all income levels, negating income alone as an explanation.

The project took almost three weeks of homework and in-class time. Students worked mainly in groups, although all wrote individual essays. I had most of the students rewrite their essays one or two times, usually because they did not adequately support arguments, and most

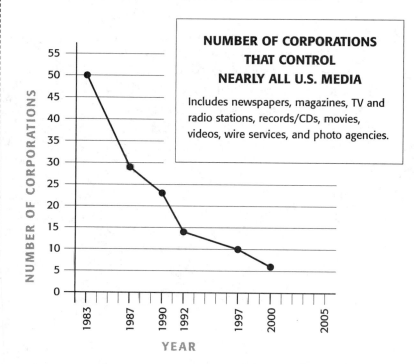

CORPORATE CONTROL OF U.S. MEDIA

LINE GRAPH TEACHING SUGGESTIONS

NUMBER OF CORPORATIONS THAT CONTROL NEARLY ALL U.S. MEDIA

Includes newspapers, magazines, TV and radio stations, records/CDs, movies, videos, wire services, and photo agencies.

Ask students to compute a best-fit equation for the line. What is the prediction for 2005? What is the mathematical problem with extrapolating this line too far into the future? How would you get around that? What do you think is the effect of the media concentration? What does this make you think about watching the nightly news?

groups reworked other parts of the project as well. We ended with two days of students reading their essays aloud and trying to arrive at individual and group understandings. Throughout the lesson, students were extremely engaged.

When I look at their projects, essays, and journals, several themes are apparent. First, many students learned that it was difficult to know if racism was a factor. Many students changed their minds, often more than once, after rethinking, further questioning and investigating, and listening to their peers' essays. I considered this important, because too often students (and adults) gravitate to simplistic "all-or-nothing" solutions and overlook real-world complexities. Most appeared comfortable with ambiguity and the realization that they needed more data to answer the question. For example, Jesse wrote:

> [Racism] is a factor because white applicants, no matter what their income was, they were always denied less times than African Americans and Latinos. And it is also a factor because the ratio of applicants denied between African Americans and whites is 5:1 and between Latinos and whites is 3:1. That data shows that racism is a factor.

> There are always two sides to a story. Racism is not a factor because we do not know whether or not those people had bad credit or were unemployed. It could be possible that a lot of those people could have been in debt. Even though the banks want to make loans they also want to make sure that they get paid.

So with the data provided it is very hard to conclude whether or not racism is a factor when it comes to obtaining a mortgage loan in the Chicago area.

Manny went back and forth several times; his final journal entry read:

> My view is that racism is a factor. First I was unsure if it was or wasn't, so I did an essay saying it wasn't but at the end changed my mind that it is. I have learned some things about mortgages. I learned that blacks have a harder time than Latinos. It's basically from really light skin to dark. Blacks have five times more trouble to get a mortgage which is a lot more times compared to whites. Latinos don't have it as bad as blacks, but not as easy as whites. My only question is: Is racism also in job applications?

FRED WRIGHT

And Nilda wrote:

Was this project to confuse us and really make us think? Because that's what it did. After our last [whole-class] discussion on Friday, everyone was talking about what we had discussed. In my first article I said that I thought racism was not a factor, after our second discussion I thought racism was a factor, but I think that we don't really know. Even though the rate for blacks was five times higher than whites in being rejected, that does not necessarily mean it is racism, it could be because of their debt, income. Or maybe it could be racism.

An interesting point Nilda and others made is that students continued the discussions outside of class, testifying to their engagement — a typical written comment was, "me and some of the girls have been talking about the conversation...."

In addition to discovering that "one question leads to another," the students also constructed alternate explanations through this process. Vanessa wondered if racism accounted for income disparity, Jesse pondered whether racism was a factor in job applications, and Nilda proposed debt and income as other possible causes of the rejection-rate disparity. Students did not base these ideas so much on the article, but rather on their own knowledge and experiences.

Common explanations were "bad credit," "less education," "worse jobs," and "no papers" (undocumented). But those explanations often led to other questions, as Carmen wrote:

I do not think racism is a factor in mortgages. At first I thought it was but not until I got to question number seven, when I defended Bank One [they had to take the bank's position in this question]. I said that if they were to ask me if racism was a factor in mortgages I would say something like this, "Racism is totally not a factor in mortgages, in fact the reason why a lot of blacks and Latinos are being rejected is because they have less collateral." In case y'all don't know what collateral is I'll try explaining it. Collateral is something you have/own that can substitute [for] money (the loan) if you can't pay — somebody asked me why whites had more collateral and I said because they have been here (in U.S.) longer. Then that one person asked me why didn't blacks have more collateral if they've been here longer, and to that question I would say that since a long time ago whites owned lots of things and blacks were not able because they were slaves.

Carmen laid bare some of the complexity but seemed to be unaware of it. She essentially

---

"AND HOW DO YOU SPEND YOUR WAGES?"

CARTOON TEACHING SUGGESTIONS

Show students the cartoon at left and ask, "What is the message of the cartoon?" Have students find the federal and state minimum wages and then calculate how much money a person who earns the minimum wage, working 40 hours per week, would have to put toward the basic necessities of food, rent, clothes, health care, transportation, etc.

Ask, "What measures could working people take in order to receive adequate wages? What role could math play in their efforts?"

named racism as a factor in African Americans lacking collateral (wealth), and hence their difficulties in securing mortgages, but did not make all the connections. A possible question to ask her might be, "Since blacks having less collateral is a factor in not getting mortgages — is racism a factor after all?" This theme, of one question leading to another, arose throughout the project, and I also promoted it by asking students to generate their own questions.

Despite my attempts, I felt that many students did not grasp the complicated sociohistorical reasons for the racial disparities in lending. Carmen came closest, but overall, I was unable to help them understand the relationships. Even though at the end of the project I gave students a handout comparing black and white income and wealth (assets) in the United States, we didn't spend enough time working through the meaning and situating it historically. It was the end of the project, right before winter break; students had been working hard and were tired, and I planned to return to it later in the year. Nevertheless, this was a weakness in the project.

During these discussions, the students' own stereotypes surfaced. Two students read aloud their essays, stating that African Americans had "dirty houses," and suggested that as a reason why they were rejected at higher rates. This sparked excellent discussions about the issues, and the general consensus emerged that one person of color cannot be seen as an example for the whole. Several students reacted, both verbally and in writing, and identified the stereotypes placed on themselves, as Latinos, as a point of unity. Laura wrote:

> That's a stereotype. I got kind of mad, because it's like people thinking Mexicans are lazy. There's this "famous" thing/picture of a Mexican guy with a sombrero hiding his face, his head on his knees, and sleeping. That's a stereotype also. It doesn't mean it has to be true.

That unity was important because it provided a basis for students to understand common oppression. This in turn can lead to students seeing themselves and others as mutual allies, creating conditions for them to act.

Making Connections

A project like this represents one way to teach mathematics for racial justice. I think it's important that students began to go beneath the surface and grapple with the cultural, economic, and historical roots of racism. The questions they asked were attempts to make sense of complicated data and a complex social phenomenon (racism) that many adults have trouble with. They did not, for the most part, fully differentiate between individual racism (e.g., a white loan officer rejecting an applicant of color because of her race) and structural racism (e.g., why African Americans have less collateral). Nor was I able to help them understand the sociohistorical context. But I believe this awareness is part of a developmental process — after all, we cannot just "tell" people an analysis, they have to construct their own interpretations. And they did begin to question and make connections.

I wanted students to raise questions, publicly discuss, and seriously study these issues. The evidence was there that they achieved these goals, while learning mathematics and using mathematics as an analytical tool with which to understand racism — even if some of their questions remained unanswered. In addition, I wanted to learn about students' perspectives and life experiences in order to be clearer about my own role in the struggles for racial justice, and, to a certain degree, most were open and willing to share their lives.

Above all, I wanted the students to have an inquiring, critical perspective. As Tita wrote, "From this project I learned that you should question everything. Like that to have a better project you should question all your answers. That's what I did." □

All students' names have been changed.

CHAPTER SIX

Sweatshop Accounting

BY LARRY STEELE

My students are out of their seats again. They're digging into backpacks, flipping down the collars of their Old Navy t-shirts and pulling off their shoes, looking for those little "Made In" tags. Made in China. Made in Vietnam.

"Hey, I was made in Vietnam," says Allan.

The students stick yellow Post-it notes on a big map of the world, marking the countries where their stuff was made. In five minutes, the Post-its obscure the coast of China, plus the Philippines, Malaysia, Indonesia, and other low-wage countries where people struggle to live on as little as $1 per day.

This isn't social studies class. It's business education. The students are examining the connections between their habits as consumers and working conditions in the countries where the goods are made.

As I teach students who hope to be future business leaders, these connections, which are often ignored in business education classes, are the central focus of our classroom work.

Today's high school students can take classes their parents never considered until college, classes like marketing, economics, and international finance. But this 21st-century curriculum needs to be taught as more than a set of technical skills. In our $31 trillion global economy, business decisions affect billions of people. That's why I think it's more important than ever for business education to connect economic skills with the values of democracy, social justice, and environmental sustainability.

That's what my accounting students are doing as they analyze the pattern of yellow Post-it notes on our classroom map. I ask them what they think the pattern shows. "Use what you've learned about accounting," I say.

Allan waves his hand. "Companies are cutting their salary expenses," he says. "They can do it by hiring people in developing countries who will work for not much money."

Allan sees the connection between the numbers on the income statement he is studying and people on the other side of the planet. That's because I supplement our textbook and its columns of worksheet numbers with readings, role plays, videos, discussions, and writing about social responsibility. As we learn how the numbers work, we also take time to look for the people behind the numbers.

A Familiar Example

My students are learning what our accounting textbook calls "the language of business." As they gain fluency, they start talking about debits and credits, business transactions, and financial reports. It's a cross-cultural language spoken by business people worldwide. The key word in this language is "profit."

"When we format a company's income statement, where do we report net profit?" I ask the class.

After some near-miss answers, Mei Li says, "It goes on the bottom line."

"Right. And what is the bottom line?" I ask her.

"It's total sales minus total expenses," she answers.

"Spoken like an accountant," I say. "But I'm looking for a broader meaning. What does the phrase mean in general. Like, 'The bottom line is you need a college education to get a middle-class job?'"

More near misses, then Mei Li ventures, "The most important thing?"

Bingo.

I want students to think critically about what can happen when profitability becomes the most important thing. Decisions based on business considerations often are presented as if they are "value-neutral" — just part of doing business. Nothing could be further from the truth. I use a lesson called "Sweatshop Math" (page 160), to illustrate the human consequences when business managers focus too narrowly on the bottom line.

To help visualize people living and working in other countries, we watch *Sweating for a T-Shirt*, a video by Medea Benjamin and Global Exchange (see Resources, page 165). Then,

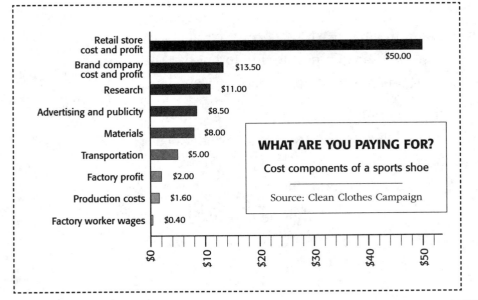

WHAT ARE YOU PAYING FOR?

Cost components of a sports shoe

Source: Clean Clothes Campaign

Retail store cost and profit	$50.00
Brand company cost and profit	$13.50
Research	$11.00
Advertising and publicity	$8.50
Materials	$8.00
Transportation	$5.00
Factory profit	$2.00
Production costs	$1.60
Factory worker wages	$0.40

the students read short vignettes from *Rethinking Globalization* about the lives of workers in low-wage countries. Using information from their reading, the students label maps of the world with summaries of local working conditions, wage rates, and hours worked. For example, one vignette tells about workers in El Salvador who "get paid 29 cents for each $140 Nike NBA shirt they sew. The drinking water at the factory is contaminated. Women raise their babies on coffee and lemonade because they can't afford milk."

As the students cut and paste labels for their maps, I help them reflect about what they know. Their ideas usually echo views they've heard before.

"If somebody takes a job its their own choice. It must be better than what they were doing before," says Rob.

"They should go on strike," says Brian.

We talk about the loss of traditional jobs, suppression of labor unions, and the pressures of global competition.

"This is making me uncomfortable," says Alicia. "Maybe I don't want to be in business if you have to take advantage of people."

"So, let's use our business skills to see if we could give those workers a better deal," I say.

We visit the Clean Clothes Campaign website, www.cleanclothes.org, to get some real-life accounting data. The site has a photo of a $100 athletic shoe labeled with all the costs that go into its retail price, including production costs, advertising, and profit for the company that owns the brand name. Soon the students are transferring the data to Excel spreadsheets and using the program to make colorful bar charts (see an example at left).

Brian is shocked. On his chart the 40-cent bar representing factory worker wages is barely visible. There's a long $13.50 bar for the brand-name company, and an even longer $50.00 bar that goes to the retail store. With such a visible

and familiar example the students immediately see options for more equitable distribution of income.

"They could pay the factory workers twice as much and it would barely dent the shoe company's share," says Brian.

We get a distorted picture when we equate success with higher production.

"Or cut advertising expense by not paying Michael Jordan so much to wear the Nike logo," adds Linh.

"Reduce the retail store's costs," says Desiree. "That's the biggest cost."

I remind Desiree that with her knowledge of fashion, she might want to work in retail sales. "Your paycheck would be part of the retail store's costs. How would you feel about taking a cut in salary?"

"Oh no! They won't be taking it out of MY pocket," she says.

Desiree gets a laugh from her classmates. But the students are beginning to see that global economic connections involve decisions that challenge their own values.

Mom and Pop vs. the Big Box Store

After looking at how financial decision-making affects people in faraway countries, I want my students to see how doing business "by the numbers" also changes lives in our own communities. As my advanced accounting students study the ownership rights of corporate shareholders, we take time to compare and contrast the interests of shareholders with the interests of local communities. In a unit I call "Mom and Pop vs. the Big Box," we read about competition between neighborhood "mom and pop" stores and the world's biggest "big box" store, Wal-Mart Corporation.

We read parts of a recent Wal-Mart annual report. It says the corporation's job "is to see how little we can charge for a product." It describes exciting career opportunities for its workforce of diverse, respectful, well-trained "associates," including 300,000 outside the United States.

> "It is a lie that our wages are adequate. Our children don't even have toys. We make baseballs out of old socks. We cannot feed our children right."
>
> **— A garment worker in Nicaragua,**
> National Labor Committee

Quang is impressed when he finds a list of Wal-Mart financial contributions to Boys' and Girls' Clubs, the United Way, and other local charities in neighborhoods near Wal-Mart stores. I ask him to create a "local citizenship" ratio to compare the amount Wal-Mart spends on community relations with its total sales for a typical store. Quang is less impressed when his calculator shows a ratio that's a tiny fraction of 1 percent.

Next we compare the company's self-portrait to one of the many critical reports about Wal-Mart's business practices around the world. As an example, here is a quote from the website of Wal-Mart Watch, www.walmartwatch.com, an organization that encourages consumers to shop at stores offering employees "good jobs with living wages:"

> There are two ways to cut costs. You can reduce waste and inefficiency. That's great. It's what makes the market system go 'round. But you can also cut costs by putting them off onto someone or something else. … You can muscle down your suppliers' prices, so they have to move production to poor communities and pay wages that won't support a decent life. You can hire part-time workers with no benefits and give them no training. Our

taxes and insurance fees will pay for their health care. … You can pressure towns for tax breaks and free roads and water lines and sewers. The other taxpayers will pick up the bill. You can pay only a fraction of the real costs of materials and energy. Nature will eat the damage. This kind of cost-cutting not only imposes injustices on others, it also undermines the market economy. It distorts prices so consumers cannot make rational decisions. It rewards bigness and power, rather than real efficiency.

This statement and Wal-Mart's annual report provide a great opportunity to "compare and contrast" texts. The Wal-Mart Watch text describes how Wal-Mart creates low prices by avoiding costs for labor, health care, and the environment. The costs don't go away, but the company leaves them out of its own accounting. It makes them "external costs," or "externalities," that will be paid later, or by someone else. Wal-Mart avoids taking responsibility for externalities. Our textbook doesn't mention this concept. It might be considered too advanced for a beginning course. I think it's pretty basic, so I introduce it here and return to the concept again and again.

The next day I hand out role-play instructions for a make-believe community meeting, including short character sketches I put together (see www.rethinkingschools.org/math for these materials). In this role play, students play Wal-Mart executives, city planning officials, shop owners, shoppers, neighborhood residents, and potential store employees.

When we're ready to begin, I shout, "Action!"

City Planner Mei Li introduces the key speaker.

Quang stands up to make his presentation to the community. He unrolls a large sheet of paper representing the plan for a 100,000-square-foot Wal-Mart Super Store.

"We want to build this great store across the street from the Martin Luther King Mall. It will sell everything you need, at low, low prices!"

My students shop at Martin Luther King Mall. They visit hair-styling boutiques owned by African Americans, a dollar store owned by a Vietnamese family, and a dozen tiny restaurants and shops.

"Hey, wait a minute," says Alicia. "My friend's aunt owns a deli sandwich shop on that street. What if all her customers start going to Wal-Mart? She'll go out of business."

"Forget those raggedy stores at King Mall," says bargain-shopper Desiree. "I want to go to one store with good prices. It's convenient."

"Yeah, and they might hire minimum-wage teenagers," says Rob.

Many of the shoppers and residents agree. Some say they've shopped at the local stores all their lives, but even they are impressed by Quang's promise of low prices for everything they need.

Later, reflecting on what we've learned during the role play, the students realize that most of their opinions were based on personal considerations. Nobody stood up and cited Wal-Mart's income statement to show the connection between low prices and low wages. Nobody argued consciously about "externalities," although some students were getting at it when they mentioned shopkeepers that might go out

HOW DO YOU LIVE...

... ON 31 CENTS PER HOUR?

You are a worker at a garment factory in the Las Mercedes Free Trade Zone in Managua, Nicaragua, sewing clothing for major U.S. labels. Your wages, including overtime, incentives, and bonuses, are about 31 cents an hour, or $14.88 per week.

Round trip bus fare to work	$2.45
Breakfast (coffee or juice only)	$1.92
Lunch (rice, beans, some cheese, soft drink)	$6.37
Rent (cost for sharing a one-room hut with another family)	$3.53
Water (from one faucet that doesn't always work)	$1.41
Subtotal (exceeds your $14.88 in weekly wages):	**$15.04**
Electricity	$1.77
Wood (for cooking on outdoor stove)	$1.88
Powdered milk (enough for two infants)	$4.08
Childcare	$6.12
School (textbooks and supplies)	$0.47
Total:	**$29.34**

Now the total is $29.34, without even including clothing, medical care, or food for dinner. The minimum wage in Nicaragua is about 25 cents an hour. How will you survive?

Source: National Labor Committee

of business if the Super Store opened. I resolve that next time, I'll find a more concrete way for students to correlate their concerns with the financial considerations affecting their community. Maybe I'll cast someone as a labor activist.

Nevertheless, the make-believe community meeting reflected real life. It was clear that by focusing narrowly on a low-cost business model, discount stores like Wal-Mart — and discount shoppers — would sacrifice important social values.

"Full-Cost" Accounting

What if the accounting system changed to include, rather than exclude, social and environmental costs? From the European Union to Japan to the U.S. Environmental Protection Agency, government agencies and innovative businesses are experimenting with ways to do this. They call it "environmental accounting" or "full-cost accounting."

I want my students to get a taste of how full-cost accounting could work, so I announce that it is snack time. We share slices of crisp, juicy apples from two paper plates labeled "local organic" and "imported." Then, I ask the students to discuss their ideas about the costs to bring each type of apple to the table.

We chart the entire life cycle of the apples, listing all the various costs involved in growing, transporting, and marketing the products. Dividing into small groups, some focusing on organic apples and some on imported varieties, the students compete to see who can make the longest, most complete list of costs. In a second column, the students mark an "I" for costs they think are included in the price of the apple and an "E" for costs they think are externalized.

I hear debates about water, organic fertilizer, and pesticides.

"Where do imported apples come from?" I ask.

"Australia!" "Mexico!" "Chile!" they answer.

"Then don't forget the fuel for long-distance transportation, and the exhaust and traffic jams," I remind them.

We consolidate the group lists on the board. The organic and imported lists look pretty similar — land, water, farm worker labor, a store. But the students begin to find differences, too.

"You can use a truck to get Washington apples to the store, but you need a ship for apples from Australia," says Ramiro.

"Organic fertilizer doesn't pollute, but non-organic ones might pollute," says Grace.

The groups think most of the costs, even highways and seaports, are included in the price of an apple.

"Diesel fuel taxes pay for highways. The store includes that in the price of the apple," says Rob.

But we find some externalities, too. Business doesn't pay for more air pollution when they burn more fuel. They don't pay if chemical pesticides get into groundwater. If companies were held fully accountable for these costs, they would have stronger reasons to adopt better practices.

Apples provide an accessible example. Full-cost accounting gets far more complicated in industries like forestry, mining, or energy. I ask for some general conclusions and trigger a short debate.

"Companies will have to pay for the damage they do instead of leaving their mess for somebody else to clean up," says one student.

"Yeah, but that will make everything more expensive," argues another.

We try to reconcile these two views and realize that the costs wouldn't change; they'd just be assigned to specific products. That would make the prices more realistic.

"If my business had to pay for all the social and environmental costs, I'd try to treat people and the planet better," says Rushawn.

Driven by the Market

Do corporate executives really want to promote social justice and environmental sustainability? Maybe there is something in the rules that stops them. The Transnational Capital Auction activity (page 152) reveals part of the answer. Student teams role-play as representatives of developing countries trying to attract foreign investment. They must bid against each other to win new factories and jobs.

As business students, my class knows what investors want: lower costs and higher profits. They quickly learn how to play the game.

"Build your factories in our country!" they say. "Our people are happy to work for $1 per day. We don't allow unions. We'll cut your taxes. Forget about environmental regulations!" Within reason, country-teams that compromise social and environmental conditions to reduce business expenses win the prize of foreign investment. After several rounds of bidding, virtually all of the countries discover that they have been competing in a "race to the bottom" — a race that destroys social and environmental values rather than promoting them.

When the game is over, we talk about why countries enter the race at all.

"Everybody needs jobs," says Irene.

"If they don't have money for their own factories, they've got to get it somewhere," says Henry.

Those are the traditional answers. But our essential question was, "Is there something in the rules that forces a competition to make things worse?"

I explain the concept of "fiduciary respon-

sibility": Corporate executives, bankers, investment advisors, and many kinds of business people are responsible for managing the money invested by other people. Generally, this means increasing the "return on investment" by increasing profits. It means people who are rich enough to have disposable income to invest get richer. People who aren't get their salaries squeezed.

"It's the bottom line again," says Mei Li.

"That's why we won the game," says Rob, whose country now allows child labor and pays starvation wages. "We cut their costs so they could make bigger profits."

There are many reasons for poverty and ecological decline. As we reflect about what we've learned from the Transnational Capital Auction, we talk about all the different ways people struggle to improve their lives. Political parties struggle for democracy against the temptation of corruption. Workers fight to organize

THE GLOBAL CAPITALIST ECONOMY

CARTOON TEACHING SUGGESTIONS

Show students this cartoon along with other Polyp cartoons (pages 12 and 143). Using data in this cartoon, have students make other cartoons and graphs to demonstrate the maldistribution of wealth in the world.

Have students do activities from *Rethinking Globalization* (see Resources, page 170) to explore the causes of this unequal distribution of wealth.

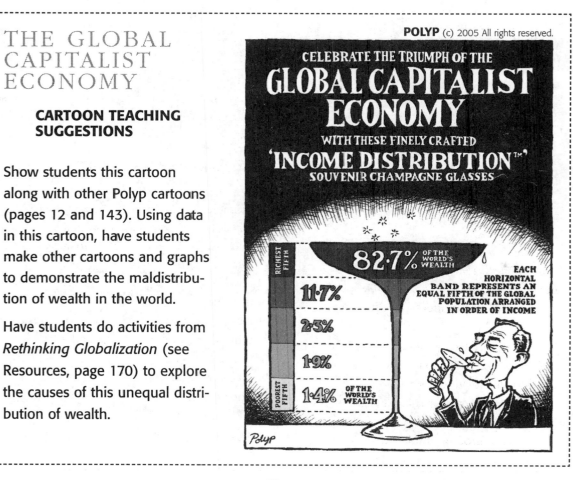

POLYP (c) 2005 All rights reserved.

unions. Teachers prepare students for more complex jobs.

But as business students, we are studying the financial rules that guide the entire game. I want the students to think critically about whether those rules are creating outcomes we truly desire.

National Accounting

The U.S. government uses a system of national accounting to measure the performance of our entire country. To understand this, I ask my students to visit the website of the U.S. Bureau of Economic Analysis, www.bea.doc.gov. They download a table that shows the historical size of U.S. gross domestic product (GDP), the dollar value of all goods and services produced in our country. Using graph paper or Excel spreadsheets, they create bar charts that show how the GDP has skyrocketed during our lifetimes to $11 trillion per year. Government leaders, economics textbooks, media commentators, and many teachers agree that the growth of the GDP means our lives are getting better and often equate "growth" with "progress."

A lesson called "What's Up with GDP?" at the website of Facing the Future, www.facingthefuture.org, illustrates the distorted picture we get by equating success with higher production. Students role-play as citizens of a make-believe town called Salmon Bay, Alaska.

Most folks in Salmon Bay earn their living by fishing, but the town also has bankers, lawyers, business owners, retail workers, and the CEO of an oil company. Henry, the banker, passes out everyone's monthly salaries in colored-paper $100 bills. We tape the bills together into chains — shorter ones for the fishing people and longer ones for the professionals and CEO. We use the total income of everyone in Salmon Bay to represent the town's GDP.

Then disaster strikes. A truck hits an oil pipeline running through Salmon Bay, the oil spills, and a thin film flows over the waters of the bay. The fish die. The fishers lose their jobs.

People get sick as oil contaminates the water supply.

But the spill boosts business for some companies.

Desiree reads an "after spill" card from the curriculum guide that tells what happens, for example, to retail business owners:

> We are sorry to say that some oil workers and fishers are out of work and are now spending less at the grocery store, movie theaters, and gas stations. However, the good news is that hotels and restaurants have been very busy since the spill, as there are many officials in town reviewing and monitoring the cleanup operations. We are experiencing a 50 percent increase in business. Doctors are busy. Lawyers are busy. The oil company CEO spends millions on the cleanup, and gets a bonus for her performance.

After the oil spill, Henry adjusts pay envelopes for everyone in town. We again measure Salmon Bay's GDP, represented by totaling everyone's colored-paper $100 bills. Allan announces the bottom line:

"The GDP INCREASED!"

This isn't what the students expected. Because traditional GDP measures only economic transactions in dollars and cents, leaving out human suffering and environmental damage, the economy of Salmon Bay looks better after the oil spill.

As always, we remind ourselves of the essential question for this class: What can we see if we look at the lives of people behind the financial numbers? We would see the wrong picture if we used only dollars and cents to account for the success of life in Salmon Bay.

In the future I plan to use more lessons from Facing the Future. They have published a useful student reading guide titled *Global Issues & Sustainable Solutions* that supports the People and the Planet curriculum. This set would help balance any business-education curriculum.

Redefining Progress

Like full-cost accounting for companies, some economists support full-cost accounting for the national economy. If we want to account for the things that truly make our lives better, what important things could we measure?

After reading about labor struggles, watching videos about environmentally sustainable development, and role-playing as economic decision-makers, my students now can list dozens of important values: "job security," "clean water," "fair pay," "health care," "safe food," "a sustainable future." I scramble to list them all on the board.

Then we visit the website of Redefining Progress, www.rprogress.org, and read about the genuine progress indicator (GPI). GPI is an alternative to the GDP index. The students discover that it balances many of the important values they listed with more traditional economic indicators. For example, the GPI includes the value of unpaid childcare and housework. It subtracts, rather than adds, the expense of cleaning up pollution.

We print out a chart showing historical growth of the GPI and lay it alongside our earlier charts of historical GDP. The traditional measure of U.S. economic success is increasing faster and faster. But when pollution, crime, natural resource exploitation, and other social and environmental factors are included, the GPI shows that our lives are improving much more slowly.

The View from the Corner Office

My students are out of their seats again. They're looking out the 40th-floor windows of a "Big Four" accounting firm, the local office of one of the world's four predominant accounting companies. From this height they can see container ships from China unloading at the Port of Seattle. They can see the homeless shelter down the street from the King County Jail. They can see the snow-capped Olympic Mountains 35 miles away across Puget Sound.

The young associate accountants who meet us are plugged-in and enthusiastic. Their clients are medium-sized businesses and large corporations. It's a people-oriented profession, they say. Everyone has a laptop computer with wireless connections to their colleagues, plus links to accounting rules and tax laws in every country of the world. They love the variety, pay, and perks of their jobs. The best news is that Big Four firms often recruit business-school students during their senior year of college. Study accounting, pass the CPA exam, and you could start your career earning $30,000 to 40,000, with great upward potential, according to the American Institute of Certified Public Accountants.

Some of my students will find a place in corporate accounting. Some may start their own accounting practices and help small businesses or individuals. Many will use their knowledge of accounting principles to be informed investors and consumers — and perhaps activists.

Whatever paths they choose, I hope they have the courage to ask the critical questions that lurk behind every financial decision: When we measure success in terms of dollars and cents, who do our decisions affect, and how? □

All students' names have been changed.

GLOBALIZATION, LABOR, AND THE ENVIRONMENT

A LINEAR PROGRAMMING UNIT

BY INDIGO ESMONDE AND JESSICA QUINDEL

Story problems often ask students to put themselves in someone else's shoes and to make decisions for them. Students are discouraged from questioning the contexts of these problems and from using their values and outside knowledge to make sense of them.

We decided to design a six-to-eight-week linear programming unit (based on the "Cookies" lesson in the Integrated Mathematics Program of the National Council of Teachers of Mathematics) that encourages students to challenge the problem context and come up with more equitable contexts of their own.

The following problem is one of the unit problems, which students work on for several weeks. Their job is to advise the CEO of a shoe manufacturing company.

sWOOsh Inc. is a new shoe manufacturer trying to introduce their product into the marketplace. They have decided to do an initial manufacturing run of two types of shoes: a basketball shoe and a skateboarding shoe. They need to decide how many of each type of shoe to make.

OPTION 1: INDONESIA

They pay workers about $750 to make 100 pairs of basketball shoes in this non-unionized factory in Indonesia and $375 to make 100 pairs of skateboarding shoes. (The workers are paid by the hour, so these are estimates based on how long it takes to make each kind of shoe.)

To make 100 pairs of basketball shoes, they need 50 square feet of synthetic material and 50 square feet of leather. To make 100 pairs of skateboarding shoes, they need 100 square feet of synthetic material.

The duties and shipping for one flat of 100 pairs of basketball shoes is the same as the cost for one flat of 100 pairs of skateboarding shoes: $700.

sWOOsh is sure that they can sell as many shoes as they produce. Their decision is limited by the following constraints: They have budgeted $15,000 for labor and $17,500 for shipping; they have 2,200 square feet of synthetic material, and 900 square feet of leather.

They want to make the most profit possible. The manufacturer's profit is $1,200 for 100 pairs of basketball shoes and $800 for 100 pairs of skateboarding shoes. How many flats of each type of shoe should they make?

As students learn about linear inequalities, we want them to also learn about global inequalities. Therefore we provide texts and videos for students to learn about globalization and the economic and environmental impact of sweatshop manufacturing.

OPTION 2: CALIFORNIA
The shoes are manufactured at a union shop in California: Labor costs increase, shipping/duties decrease, materials are the same, and profits decrease. Students start to understand how the choice of different amounts for the variables influences the outcomes for the situation.

EXPLORING FURTHER
Above is just one example of a problem that students explore in this unit. Other problems that students explore help them to understand the mathematics of linear programming and to learn more about economics and globalization. We provide text and video resources for students to do background research into these topics. For the culminating project in this unit, we organize a debate in which students make recommendations about where the shoes should be produced and under what conditions. Students represent environmentalists, domestic labor, Indonesian labor groups, and corporate interests and are required to use mathematical evidence in their arguments. ∎

POVERTY AND WORLD WEALTH

RECOGNIZING INEQUALITY

BY SUSAN HERSH AND BOB PETERSON

An important part of a person's understanding of global issues is the recognition of the dramatic inequalities between nations and social classes within countries. Math is an essential tool for acquiring this understanding.

The purpose of this activity is to demonstrate graphically the vast differences in wealth between different areas of the world. It combines math, geography, writing, and social studies.

We remind students of some of the things we learned about colonialism, such as how great quantities of silver and gold were stolen from the Americas and taken to Europe. We also explain that current relations between countries and international organizations, such as the World Trade Organization, also affect how much wealth countries possess. We make sure that students know the following terms: resources, GNP, wealth, distribution, income, power, and colonialism. (Additional teaching ideas that help set the context for this lesson can be found in *Rethinking Globalization*. See Resources, page 170.)

MATERIALS
- 11" x 17" world maps for each student or pair of students.
- 50 chips (25 of one color and 25 of another) for each map.
- 25 slips of paper with "I was born in [name of continent, based on chart]."
- 25 chocolate chip cookies.
- World map laid out on playground, or signs with names of continents and yarn to distinguish boundaries.
- Transparency of "World Population and Wealth" table (page 67).
- Six "negotiator" signs with yarn to hang around students' necks.
- Writing paper and pens or pencils.
- Additional cookies for students who don't get any during the simulation (optional).
- Worksheets (optional) for students to write down their estimates (available at www.rethinkingschools.org/math).

SUGGESTED PROCEDURE

1. Give each student or pair of students a world map. Have them identify the continents and other places you have been studying.

2. Ask students how many people they think are in the world. After students have guessed, show them an almanac or a website with a current estimate. Ask: If we represent all the people in the world with 25 chips, how many people is each chip worth? (For 6.5 billion people, for example, each chip would represent 260 million people.)

3. Give 25 chips to each student/group and have them stack them on the continents, based on where they think people live. Have students write down their estimates using the worksheets mentioned in the materials list or on a piece of paper. Discuss student estimates and then tell them the accurate figures. Have them rearrange their chips to reflect the facts. Ask students what the differing stacks of chips tell them about the world's population.

4. Explain that you are now going to give them another 25 chips of a different color and that they represent all the wealth produced in the world (the monetary worth of all the goods and services produced every year, from health care to automobiles). Each chip therefore represents 1/25 of the world's total amount of goods and services produced. Tell the students to put the chips on the continents to indicate their estimate of who gets this wealth.

5. Discuss student estimates and record them on the chalkboard. Have students reflect on the sizes of the two different sets of chip stacks, representing population and resources. Collect the chips.

6. Tell students you are going to demonstrate how population and wealth are distributed by continent. Have each student pick an "I was born in..." slip from a container labeled "chance of birth." Students may not trade slips. (As you distribute the slips, listen for stereotypical reactions to the continents — these will be useful in the follow-up discussion and will indicate possibilities for future lessons.)

7. Have students go to an area that you have designated to represent that continent. (Playground maps work great for this.) After students are in their areas, remind them that they each represent about 260 million people and that you are going to distribute the world's wealth. Have each continent/group designate one person to be a "traveling negotiator" and distribute a traveling negotiator sign to those people.

8. Explain that once the bag of resources is passed out to a representative from each continent, each group needs to sit in a circle and discuss their situation. Tell the students there will be

a cross-continent negotiation session, then a time for the traveling negotiators to return to their home base to discuss their negotiations with the rest of their group, and finally a time for any trading or donating of resources. Students on each continent are to talk about how many resources they have compared to people of other continents and to discuss ways they might negotiate to

Teachers can connect students' feelings about fairness to the data on world wealth.

increase their resources. They may plead and/or promise. (Note: Every continent, except North America, will have at least one "stay-at-home negotiator" and one traveling negotiator. The North American person can stay put or travel throughout the world. Also note that because Mexico is by most definitions part of North America, I explain to the students that for the purpose of this simulation we will be using "Latin America," which includes South America, Central America, and Mexico; instead of "North America," we will use "United States/Canada.")

9. Use a popular treat — such as chocolate chip cookies — and distribute them according to the percentages noted in the chart. Announce the number of treats you are giving to each continent as you do so. Provide a paper bag for each continent to keep their treats in. As you dramatically place each of the resources into the bag, remind students they are not to eat the treats until after the negotiation session.

10. Announce that the negotiation session is to

begin. Only traveling negotiators may move to a different continent. When they come, they should sit in a circle with the stay-at-home negotiators and discuss the distribution of wealth and what should be done about it.

11. After five or 10 minutes, tell all traveling negotiators to return to their home continents. Each group should then discuss the negotiations. After a few minutes, announce that the trading session may begin, and if a continent wishes to trade or donate resources, they may. After that, instruct the people holding the resource bags to distribute the resources to people in their group.

12. Give each continental group tag board and markers. Tell them to make some signs that describe what they think of the way the resources were distributed.

13. Bring students back together for a whole-class discussion. Have each group share their posters and perspectives. Show students the information from the world wealth chart via a transparency or handout. Connect their emotions and feelings of fairness to the information on the chart. (At this time, a teacher might give out additional treats to those students who did not get any.)

QUESTIONS WORTH POSING IF THE STUDENTS DON'T ASK THEM THEMSELVES

• How did the distribution of wealth get to be so unequal?
• What does the inequality of wealth mean in terms of the kinds of lives people lead?
• Who do you think decides how wealth is distributed?
• Should wealth be distributed equally?
• Do you think that, within a particular continent or nation, wealth is distributed fairly?

• How does the unequal distribution of wealth affect the power that groups of people hold?
• Within our community, is wealth distributed fairly?
• What can be done about the unequal way wealth is distributed?
• Who can we talk with to find out more information about these matters?

If your students have studied colonialism, ask them what role they think colonialism played in creating this inequality.

After the discussion, have students write an essay about their feelings, what they learned, what questions they continue to have, and what they might want to do about world poverty. Some students might also make wall posters that graphically depict the inequality of wealth.

FOLLOW-UP ACTIVITIES

A few days after this simulation, "Ten Chairs of Inequality" (page 135) is a useful activity to help students understand that wealth is also unequally distributed in individual countries.

Students also can do follow-up research on related topics, such as: the role colonialism played in the wealth disparity; how current policies of U.S. corporations and the U.S. government affect people in poorer nations; the role of groups such as the WTO and the International Monetary Fund; and what different organizations and politicians are doing about world poverty. (Refer to the list of "Organizations and Websites for Global Justice" in *Rethinking Globalization*. See Resources, page 170.)

Three notes of caution with this activity: First, as with any simulation (or role play) this should be understood to be just that — a simulation. We can in no way reenact the violence of poverty and hunger that kills tens of thousands of children daily. We are providing a mere glimpse. Second, while Africa and other areas south of the equator do not have lots of wealth as defined by GNP, those areas have great human and natural resources and this fact should not be lost on the students. Finally, in this simulation we seek to describe rather than to explain current power and wealth arrangements. They can, however, be powerful tools in motivating students to want to figure out the answer to the essential question: Why? ■

WORLD POPULATION AND WEALTH

DATA TABLE

CONTINENT	POPULATION (in millions)	% OF WORLD POPULATION	# OF STUDENTS (in a class of 25)	WEALTH (GNP) (in billions of dollars)	% OF WORLD GNP	# OF TREATS (out of 25)
Africa	906	14.0	4	495.4	1.8	0.5
Asia	3,905	60.4	15	7,172.6	25.5	6
Oceania	33	0.5	0	442.4	1.6	0
Europe	728	11.3	3	9,606.3	34.2	9
U.S. and Canada	331	5.1	1	8,933.6	31.8	8
Latin America	561	8.7	2	1,430.7	5.1	1
WORLD TOTAL	6,465	100.0	25	28,081.0	100.0	24.5

Sources: World population figures are from the U.N. Population Division, www.un.org/esa/population, based on estimates from July 2005. GNP figures are from the World Bank, quoted in the *Universal Almanac,* 1994. GNP is defined as the total national output of goods and services. Percentage of world wealth is an estimate based on total GNP (not shown in the table).

For purposes of this chart, one third of Russia's GNP is attributed to Asia and two thirds to Europe. Latin America includes Mexico, the Caribbean Islands, and South America. The data were broken down this way to highlight the great disparity of wealth between Mexico, on the one hand, and the United States and Canada, on the other.

UNEQUAL DISTRIBUTION OF WEALTH IN THE UNITED STATES

RECOGNIZING INEQUALITY, PART TWO

BY MICHAEL LANGYEL

Display the graphic "U.S. Households and Wealth" (at right). Explain to students that there is a difference between income and wealth. Economist Edward Wolff, author of *Top Heavy: A Study of the Increasing Inequality of Wealth in America* argues that to analyze economic inequalities, one must go beyond the annual income of a person or family and look at their wealth, which is defined as the dollar value of assets, minus any debts or liabilities, held by a household at any one time. (Income, by comparison, refers to the flow of dollars over a period of time, usually a year.)

Ask the students what the house graphics represent and what the money-bag graphics represent. Ask: "What do these data show us?"

Tell students that they are to represent this data in three-dimensional form using 100 pennies (or chips) and a chart of 100 squares.

Have the students draw out a 10 x 10 chart with squares large enough that a penny can fit entirely inside a single square. (For a downloadable PDF of a chart like this, see www.rethinkingschools.org/math.)

Tell the students that each square represents one percent of the households in the United States, and each penny represents one percent of the country's wealth.

Have the students work individually or in groups of two or three to construct their three-dimensional graph — stacking pennies or chips on squares to represent the distribution of wealth among different segments of the U.S. population.

ADDITIONAL PROCEDURE

For students needing more explicit instructions, suggest the following:

1. First divide the 10 x 10 grid in three regions by coloring each with different colors:

    - 1 square = richest 1%
    - 19 squares = the next richest 19%
    - 80 squares: the remaining 80% of the families.

2. Next divide the pennies into three piles: 39 pennies for the richest square, 46 pennies for the next region, and 15 pennies for the third region.

Note: Some students may have difficulty stacking 39 pennies on the richest square. Also some will struggle to divide 15 pennies among the 80 squares in the largest region.

QUESTIONS

Have students reflect in writing or in discussion on these questions:

1. What do these data tell us about the distribution of wealth in the United States?

2. Why do you think wealth is distributed this way?

3. What is the relationship between wealth and poverty?

4. What government policies on spending and taxes do you think the three different groups might support or oppose? What other data might be useful to more deeply understand the distribution of wealth in the United States? ■

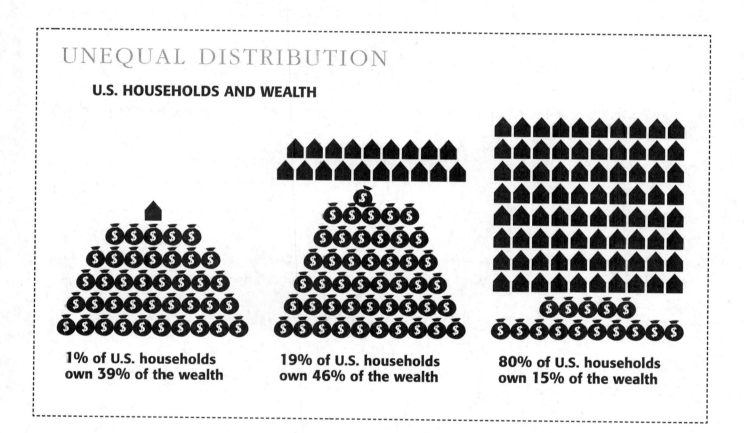

UNEQUAL DISTRIBUTION

U.S. HOUSEHOLDS AND WEALTH

1% of U.S. households own 39% of the wealth

19% of U.S. households own 46% of the wealth

80% of U.S. households own 15% of the wealth

CHAPTER SEVEN

Chicanos Have
Math in Their Blood

Pre-Columbian Mathematics

BY LUIS ORTIZ-FRANCO

Mathematics education dates its beginnings to the time when human beings began to quantify the objects and phenomena in their lives. Although the process of counting (one, two, three...) was the same for different groups of people around the world, the symbols by which they represented specific quantities varied according to their own particular cultural conventions. Thus, the Babylonians, Romans, Hindus, Egyptians, Angolans, Chinese, Aztecs, Incas, Mayas, and other groups each wrote numbers differently.

Likewise, cultures that achieved a level of mathematical sophistication that allowed them to manipulate their number symbols to add, subtract, multiply, and divide and to perform other algorithms did so in different ways. Today, even within a single society, various groups of people (for example, accountants, physicists, engineers, mathematicians, chemists, and so on) view and manipulate mathematical quantities differently from one another. The study of the particular way that specific cultural or ethno groups — whether they are different national, ethnic, linguistic, age, or occupational groups or subgroups — go about the tasks of classifying, ordering, counting, measuring, and otherwise mathematizing their environment is called ethnomathematics.

The ethnomathematics of pre-Columbian cultures is a topic frequently overlooked in discussions about the cultural achievements of pre-Columbian civilizations, and omitted from college-level textbooks on the history of mathematics. It is important that we focus on pre-Columbian mathematics: Such a focus broadens our perspectives on pre-Columbian cultures and may stimulate us to integrate new perspectives and topics into our classroom teaching. The ultimate beneficiaries of these educational practices will be North American society in general and North American children in particular.

The integration into school mathematics of pre-Columbian mathematics is important for both political and mathematical reasons. The teaching of the mathematical traditions of pre-Columbian cultures can contribute to achieving a crucial political goal: infusing multiculturalism into education. Students will thereby develop an appreciation for the diverse ways different cultures understand and perform mathematical tasks. This will expose students to the sophisticated mathematical traditions of other cultures and demonstrate that performing mathematics is a universal human activity.

For Chicano students in particular, studying pre-Columbian mathematics will allow them to learn more about their ancestors in both cultural and mathematical contexts. This integrated approach can do much to instill pride in their culture and also increase their confidence in their ability to learn and do mathematics and, perhaps, later participate in mathematics-based careers.

Chicanos are mestizo, a blend of European and Mexican Indian ancestry. In this country, the cultural roots of Chicanos in pre-Columbian cultures are acknowledged at the social level but are usually ignored in the American educational system. In a scene in the movie *Stand and Deliver*, teacher Jaime Escalante attempts to motivate his Chicano mathematics students at Garfield High School in East Los Angeles by

Chicanos' impressive history of cultural achievement has been almost entirely ignored in U.S. schools.

saying: "You burros have math in your blood." Escalante's comment surely will seem sadly ironic to readers familiar with the statistics regarding the low educational achievement of many Chicano students, and familiar with both the history of mathematics in Maya and Aztec societies and the relatively minuscule number of Chicanos who pursue mathematics-based careers in the United States.

Despite a long and distinguished heritage in the sciences, arts, and letters in their own culture, Chicanos are one of the least educated groups in this country. Their impressive history of cultural achievement has been almost entirely ignored in U.S. schools for at least two reasons. First, the Western orientation of the educational process largely disregards the achievements of conquered indigenous civilizations and their descendants, such as Mexican people. Second, as a result of this ethnocentric orientation, many teachers and other school officials in the United States are unaware of the mathematical accomplishments of pre-Columbian societies.

This article discusses a pre-Columbian number system that played an important role in the cultural activities — such as commerce and dating historical events — and the development of Mesoamerica. (The term "Mesoamerica refers to the geographical region that encompasses the area from northern-central Mexico to northern Costa Rica.) In this essay, I discuss the Mesoamerican number system and the origins of the vigesimal number system, and make some instructional suggestions.

The Mesoamerican Number System

The Mesoamerican number system is a positional vigesimal (that is, base 20) system. It employs only three symbols to write any whole number from zero to whatever quantity is desired. This symbol represents 0:

The dot represents the quantity of 1: •
And the horizontal bar represents the quantity of 5: ▬▬

To write the numbers 0 to 19 in this system, the two processes of grouping and addition are used. Numbers 2 to 4 are written using an addition process. For the number 5, five dots are grouped into a horizontal bar. Numbers 6 to 19 are written using the addition process (see Figure 1).

For numbers larger than 19, a vertical positional convention is used (see Figure 2). In this convention, the bottom level is for the units, the next level up is for the 20s, the third level up is for the 400s (1 x 20 x 20), the fourth level up is for the 8,000s (1 x 20 x 20 x 20), and so on in powers of 20. For instance to write 20, we write the symbol for 0 in the first level and a dot in the second level. To write 65, we write three dots in the second level (3 x 20 = 60) and a horizontal bar, representing five units, in the first level.

This vigesimal system of numeration is practical to use and can easily be adapted to classroom instruction. For example, the operations of addition and subtraction are relatively straightforward processes. In the case of addition, one has to remember that since 20 units in a lower level are equivalent to one unit in the next level up, 20 units in a lower level are replaced with one dot in the level above it. Figure 3 shows the sum of 8,095 plus 1,166, before and after the grouping process.

This addition example points to rich mathematical experiences in which students can be engaged. Unfortunately, beyond addition, subtraction, and associated algorithms, we do not know whether the Mesoamerican civilizations knew how to multiply, divide, or perform other mathematical algorithms with this vigesimal system. But we do know that the Mayas wrote books on paper just as we do. We know that for 1600 years before Columbus accidentally arrived in the "New World," the Mayas wrote and kept thousands of books in which they recorded their history and cultural achievements.

Tragically, however, the Spanish conquerors and missionaries burned and otherwise destroyed all of the Mayan libraries and archives. Possibly some of those destroyed books contained information on algorithms and other mathematical systems that pre-Columbian societies devised. We know that Mayan astronomers had calculated the cycles of the heavens so exactly that they could predict solar and lunar eclipses to the day, hundreds of years in advance. For example, a Mayan astronomer predicted, some 1200 years in advance, the solar eclipse that occurred on July 11, 1991.

The Mayas knew the synodical revolution of Venus, and some scholars argue that the Mayas also knew the synodical period of Mars and perhaps had parallel knowledge about Mercury, Jupiter, and Saturn. Given their ability to make these calculations as well as to predict celestial phenomena, it is reasonable to believe that they knew how to perform mathematical algorithms other than addition and subtraction. This belief is rooted in the origins and uses of their vigesimal number systems.

Origins of the Vigesimal Number System

Archaeologists and other scholars maintain that humans first inhabited North America around 30,000 years ago and, in particular, Mexico about 9,000 years later. Groups of hunters and gatherers roamed Mesoamerica for thousands of years before they became sedentary. Jacques Soustelle pins down the advent of agriculture in Mesoamerica at approximately 4000 B.C. However, the organized life that can be called civilization in the region began approximately 5,000 years ago. The social evolution of Mesoamerica can be traced from the hunter-gatherers through the successive civilizations of the Olmecs, Zapotecs, Mayas, Toltecs, Aztecs, and so forth.

The earliest evidence of numerical inscriptions that used positional systems of bars and dots has been traced to the Olmecs in approximately 1,200 B.C. This date is significant, since some 800 years before Aristotle, Plato, and Euclid (whose society did not have a positional number system) began making contributions to Western culture, the Olmecs were already using a positional system. It is worth noting that it was not until 499 A.D. that the Hindu-Arabic number notation using zero in a positional convention first occurred.

The Zapotecs of Oaxaca used the Mesoamerican vigesimal system in their calendars between 900 and 400 B.C. (Between 400 B.C. and 300 B.C., the Izapan culture used the same convention. Later, the Mayas, to whom the vigesimal system is mistakenly attributed, used this system extensively, between 199 A.D. and 900 A.D. The Mayas developed their amazingly complex calendar system and astronomical sciences around this mathematical system hundreds of years before the achievements of Galileo and Copernicus.

Figure 1: MESOAMERICAN NUMBERS 0–19.

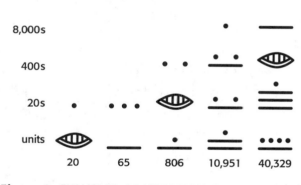

Figure 2: EXAMPLES OF MESOAMERICAN NUMBERS BEYOND 19.

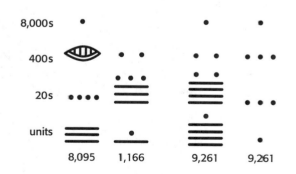

Figure 3: THE SUM OF 8,095 PLUS 1,166, SHOWING THE RESULT BEFORE AND AFTER REGROUPING.

Recommendations to Teachers

The pre-Columbian positional number system can be taught at various educational levels. As we have already seen, it could be included in the elementary school curriculum as a way to deepen how students understand the decimal system. In fact, the Mesoamerican system may be easier for children to grasp than the decimal system: The vigesimal system is visual, and the representation of quantities involves only three symbols — 0, 1, and 5 — and manipulative materials can be adapted to give them physical representation.

For example, Dienes blocks can be adapted to the Mesoamerican base 20 system. Teachers can assign to the smallest blocks the value 1, to the intermediate-sized blocks the value 5, and to the larger blocks or to a group of four intermediate-sized blocks the value 20. Alternatively, in classrooms that do not have Dienes blocks but have manipulative materials of different colors and sizes (rods or chips), different colors or sizes of rods or chips can be used for 1, 5, and 20.

At the middle school and high school levels, discussion of this number system can be included in social studies classes as well as mathematics classes, to broaden student appreciation of the cultural achievements of ancient peoples and the fate of conquered civilizations. Teachers can use this topic to illustrate that impressive mathematical achievements of Mesoamerican civilizations were ignored, devalued, or destroyed as part of the rationale for subjugation and domination. For presentations in social studies courses, a map of Mesoamerica is indispensable and can be obtained from the National Geographic

---

ACTIVITY BOX

THE HIDDEN GRAIN IN MEAT

BY STEPHANIE KEMPF

One billion of the world's people do not get enough to eat, yet half the grain grown in the world is fed to livestock. Why? To fatten the cattle up for sale to people who can afford to buy meat. Chronically hungry people rarely have the money to buy meat.

Most cattle today do not graze freely on pasture grasses — if they did, their meat would be leaner and healthier. Instead, they are penned up in crowded "feedlots" and given large quantities of grain. The meat from grain-fed cattle is higher in fat.

For every 16 pounds of grain fed to a cow, we get only one pound back in meat on our plates. Producing that pound of meat requires 2,500 gallons of water. In many areas of the world, people do not have access to even a small amount of clean drinking water and must walk miles a day to get it.

DO THE MATH

If your entire class went to McDonald's and each student ate one Quarter-Pounder, how much grain was used to produce the class's lunch? How much water was used?

Explain why you think this is or is not a problem. If it is a problem, what are possible solutions?

From *Finding Solutions to Hunger: Kids Can Make a Difference*. See Resources, page 171.

Society (for additional sources, see Resources, page 173). Teachers can consult other references for more details on the historical origins and uses of this numerical system and for more information on the pre-Columbian cultures who used it.

In mathematics classes at the middle school and high school levels, students can explore interesting mathematics through the Mesoamerican vigesimal system. Teachers can devise exercises comparing the polynomial representation of numbers in our decimal system and the vigesimal system. For example, students can explore interesting mathematics by relating powers of 10 to the value of digits in numerals in the decimal number system and the value of units in the vigesimal system: While the value of a digit in the decimal system is multiplied by a power of 10 that corresponds to the place of the digit of the numeral, the value of the same number of units in the same corresponding place in the vigesimal system is multiplied by a power of 20. This can lead to discussions about powers of 20 as a product of powers of two and powers of 10, to illustrate that the value of units in the vigesimal system increases exponentially faster than the value of the corresponding units in the decimal system. This in turn can serve as a natural introduction to topics related to exponential growth and exponential functions.

Furthermore, in mathematics classes where students are already proficient with the usual algorithm for multiplication in the decimal system (in grades five through 12), teachers can also include classroom activities or homework assignments requiring students to use their creativity when working with the vigesimal system. For instance, the teacher could break the class into groups of three or four students each and ask the groups to generate ideas of how to carry out multiplication in this number system. This idea can be extended to include division as well. These challenging assignments may turn into group projects that can last for an extended period of time.

Given the sophisticated system of ancient Mesoamerican mathematics and the gross under-representation of Chicanos in mathematics-based careers in the United States, the comment of Jaime Escalante to his students is indeed sardonic. The legacy of racist discrimination against the cultures and native peoples of Mesoamerica, which resulted from the military conquests and colonization ushered in by Columbus's arrival in the "New World," has continued to this day in the imperialistic practices of U.S. society. It has resulted in an educational system in this country which effectively ignores the rich tradition of excellence in mathematics in the Chicano students' background and fails to instill in young Chicanos a sense of pride in their heritage and a positive self-image.

Despite all this, some modern Chicano mathematicians have made valuable contributions to applied and abstract mathematics — David Sánchez, Richard Griego, Manuel Berriozabal, Richard Tapia, and Bill Velez, to name a few. Their contributions should be used to encourage Chicano students to pursue the exceptional mathematical heritage of their pre-Columbian ancestors. □

This article was adapted from a version which first appeared in *Radical Teacher* magazine. For a link to all references and citations included with the original article, see www.rethinkingschools.org/math.

CHAPTER EIGHT

Understanding Large Numbers

BY BOB PETERSON

"It's a zillion!" calls out one student in response to my question of how to read the number "1,000,000,000" I've written on the overhead. "No, it's not, it's a million," argues another.

Despite the fact that my fifth graders have been taught place value throughout their elementary years, there is something about big numbers that lends itself to guessing. Perhaps it's the omnipresent state lottery advertisements that tend to blur big numbers together. Or more likely it's the fact that big numbers are just difficult to read, much less understand. Imagining a billion boggles my mind, whether I'm trying to fathom the number of galaxies swirling around the universe or the number of molecules in a drop of water.

Kids are fascinated by big numbers, especially if they connect with the real world. Thanks to the U.S. government's addiction to military spending, students have an endless stream of large numbers to study. And those numbers just get bigger and bigger!

The growth in the military budget comes as schools face massive budget cuts. Teaching about these matters provides students an opportunity to improve their understanding of large numbers and, more importantly, to understand the power of math in debates about the future of our communities and world.

Before I delve into budget issues with my students, I do a couple of activities to help them put meaning behind place value. One year, the night before I was to start my mini-unit on big numbers and budgets, the students' homework was to ask family members what they thought one million and one billion meant. The next day the students shared their responses, which ranged from the precise, "One million is one thousand thousands;" to the comical, "It's what you get when you win the lottery;" to the practical, "It's enough money to buy everything we ever need and still have some left over."

I asked the students, "How many days equals a million seconds?" After some initial guesses, the students worked in groups with calculators using different strategies to solve the problem. Eventually they came up with about 11.6 days.

I then asked how long it would take for us to count to a million. Some students suggested we just do it and time ourselves. Others were more skeptical. After some practice with six-digit numbers we estimated that it would take, on the average, about two seconds a number. Some more calcuation and the class realized it would take a little over 23 days. "I'm not going to be wasting my time doing that," one student proclaimed.

To visualize a million, I asked the students to look closely at a strand of their hair and then told them that a million of those piled on top of each other would reach up to a seven-story building. I also showed the students the book *How Much Is a Million?* by David Schwartz.

Some of the pages are filled with tiny stars — 14,364 per page. The book encourages students to guess how many pages of stars it would take to reach a million, and they are surprised to find it would take 70 pages.

I then repeated some of these activities with a billion. The students soon discovered that their calculators did not go that high, so we did some whole-class work. We calculated that one

Students can learn to understand the power of math in debates about the future of our world.

billion seconds equals about 32 years. After timing the counting of a series of very large numbers, we estimated it would take about three seconds a number to count to a billion — leading us to conclude it would take almost 96 years to count to a billion. We examined the star-filled pages of *How Much Is a Million?* and calculated that it would take 70,000 such pages!

Billions for War, Millions for Schools

Next I wrote $80,000,000,000 on the chalk board and had a student read it. I then wrote the number $10,000,000 and did the same. I had students guess the significance of those numbers and then explained that the former is the estimated cost of war and occupation of Iraq for a year, and the latter is approximately the amount of money that is going to be spent renovating and adding to our 100-year-old building. I asked, "How many schools like ours could get a major renovation for the cost of just one year of the Iraqi war?" After a wide range of guesses, I asked how we could know for sure.

The class decided to go around the room counting by 10 million and that Markese would keep track of the number of school construction projects that could be bought. As each student added another 10 million, Markese made a mark on a piece of paper. After we went around the class two times, we had only reached 460 million and it was clear our effort to count to 80 billion was going to take a while. When we finally reached one billion, Markese announced we could rebuild 100 schools. I stopped the counting and suggested the students use mental math to figure out the final solution. "8,000!" one student called out. "That's a lot of schools that could be rebuilt!" added another. "That's more than in all of Milwaukee!" added a third. In fact, a quick check with the State Department of Public Instruction showed that there are just over 2,100 public schools in all of Wisconsin.

Later, as part of this mini-unit, the students discussed graphs from United for a Fair Economy (UFE) that contrast the U.S. military budget with federal social spending and with the military budgets of other countries around the world. Using UFE's data (see graph, at right), students figured out that one Stealth Bomber, at the cost of $2.1 billion, could have paid for the annual salary/benefits of 38,000 teachers. This was of extra significance to my students because our school principal had just informed their parents we'd be losing two teachers (gym and music), half our librarian time, and two paraprofessionals. "Just for a little part of one of those bombers, we could have all our teachers back," one student said.

I did this unit at the very end of the school year, in the midst of news of mounting budget cuts and continuing conflict in Iraq. I wanted to give my departing fifth graders a different perspective on the cuts: Too often the talk about budgets is filled with hopelessness and inevitability. I knew that in just a few days we'd only touched the surface of a complicated issue. I resolved that during the next school year I would cover this earlier, so that we could spend more time looking at how policymakers make budget decisions, the relative merits of various types of spending, and what social action groups are doing on these issues.

Holding students' attention during the waning days of the school year is always a challenge. This lesson held their attention, and it was a fitting way to end the school year as the students prepared to go to middle schools that also faced drastic cuts.

"This really isn't fair," one student wrote when I asked the class to reflect on the matter in their journals. "So much money is being spent when our schools need so little."

Actually our school — like most schools — needs a lot. It's just that the amount is so little compared to the hundreds of billions of dollars this nation spends each year on the military budget.

Hmmm, I wonder how many schools that could rebuild … and how many jobs that would create? □

FOR MORE INFORMATION ON MILITARY SPENDING

ORGANIZATIONS AND THEIR WEBSITES

Center for Defense Information, www.cdi.org

Center for Popular Economics, www.populareconomics.org

Center of Budget and Policy Priorities, www.cbpp.org

Children's Defense Fund, www.childrensdefense.org

Cost of War, www.costofwar.com

Dollars and Sense, www.dollarsandsense.org

New American Century, www.newamericancentury.org

National Priorities Project, www.natprior.org

United for a Fair Economy, www.ufenet.org

War Resisters League, www.warresisters.org

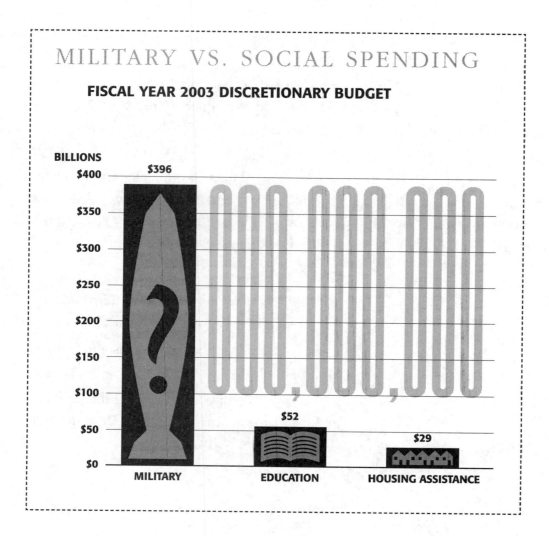

MILITARY VS. SOCIAL SPENDING

FISCAL YEAR 2003 DISCRETIONARY BUDGET

ACTIVITY BOX

HOW MANY SEARS TOWERS TALL IS $135 BILLION?

BY ERIC GUTSTEIN

In 2006, Congressional appropriations for the Iraq War will reach $315 billion. A lot of money, no? But how much really is that? Imagine that you had brand new crisp $1 bills, and also imagine that you could stack them on top of each other and they would not fall over. Suppose you stacked up all 135 billion of them.

Now think of the Sears Tower in downtown Chicago. At about 1450 feet high, it's the tallest building in North America and one of the tallest buildings in the world. Imagine yourself standing at the bottom of that building, looking up, trying to see the very top. Next think about your stack of $1 bills, the 135 billion. Would they reach as high as the Sears Tower? Half as high? Twice as high? More?

DO THE MATH

1. Write down how tall you think $135 billion is without doing any calculations.

2. Now do the calculations and figure it out.

3. Complete questions 2 through 6 from the activity "How Much Does the War Cost?" (page 29).

CHAPTER NINE

"With Math, It's Like You Have More Defense"

Students Investigate Overcrowding at Their School

BY ERIN E. TURNER AND
BEATRIZ T. FONT STRAWHUN

In an overcrowded New York middle school, students discovered that math was a path to investigating and working to change conditions at their school.

After planning a unit with Erin, a university professor, Beatriz taught a six-week unit where students used mathematics to investigate issues of overcrowding at Francis Middle School, which is located in a predominantly working class African-American, Dominican, and Puerto Rican community in New York City. Erin was present in the classroom for all of the class discussions.

When the school opened in 1990, following a district call for the creation of more "small schools" for middle-grade students, the founding (and current) principal faced the near impossible task of finding available space for the school within the district. At that time, the elementary school that currently shares a building with Francis had vacated the top floor because of continual problems with a leaking roof. Francis' principal claimed the space for the new school, repairs began, and the school opened shortly thereafter.

Although Francis began as a small school, districtwide changes in student enrollment caused the school's student population to grow dramatically, from approximately 145 students to 213 students, and it was projected to grow another 15 to 20 percent (an additional 30 to 40 students) the following year. So as students climbed the five flights of stairs to reach the once pigeon-infested fifth floor that now housed their school, they worried about how the narrow halls and lack of classroom space might accommodate their already cramped school community. Some students were concerned about the potential fire hazards created by the school's long, narrow hallways. A building across the street had recently been damaged in a fire, and in the aftermath of 9/11, students were all too aware of the dangers of being trapped in a burning building. Others felt their classrooms, several of which had recently been subdivided to create multiple rooms, were too small and overcrowded. Floor-to-ceiling columns and smaller poles scattered throughout classrooms made simple tasks like seeing the chalkboard difficult.

It was students' concerns about the lack of space at their school that guided the initial development of this project. One of our primary goals was to design a unit of study that drew upon students' interests and experiences and provided students opportunities to learn and use mathematics in personally and socially meaningful ways. With this in mind, Beatriz asked students to list issues about the school and local community that concerned them.

Several topics appeared repeatedly in students' lists, including violence in the neighborhood; health issues such as AIDS, racism and sexism in the media; and the "space crisis" at the school. While any one of these topics might have sparked a rich mathematical unit, we selected the issue of "Overcrowding at Our School" for several reasons, including (a) the rich mathematical content the unit would draw upon, (b) the opportunities it would provide for students to generate their own data versus analyzing data from an external source, (c) the salience this issue had for students, and (d) the potential links to issues of equity and fairness.

Initially, students claimed they were more crowded than other schools and were eager to speak out in hopes of increasing their school space. Students were particularly bothered by the disparity they observed between their own school space and that of Longmore, another small middle school that had recently moved into the fourth floor of the same school building. (Note: When the fourth floor initially became available, Francis lobbied to move into it, but their request was not granted, and the space was instead awarded to Longmore, a technology magnet school that attracted affluent, predominately white families from across the district.)

Yet they were not sure how to talk about the crowding in terms that might convince others, and it was unclear to them how mathematics could support their argument. To help students connect their concerns about overcrowding with mathematical tools that would support their investigation, Beatriz posed questions such as: How can we show them how much space we have? What kind of information would we need to collect? What kinds of measurements should we make? How can we prove that we are more crowded than Longmore? Students quickly realized that quantifying the school's space would be helpful.

One student, Jhana, raised a concern about the tight space in the hallways after second period, a time when all 213 students in the school were simultaneously released from their classrooms. In class, she repeatedly lobbied to investigate this. "What we need to know," she argued during a discussion, "is after second period, because that's when the most kids come out. How many kids get dismissed? ... And [we need] the area of all the hallways." Other students agreed that finding the area of classroom and hallway spaces would "give proof" to their claims of overcrowding. So Beatriz prepared a series of mini-lessons that addressed concepts such as linear measurement and how to find the area of spaces with mixed dimensions, such as a hallway that measured 10-1/2 meters by 1-1/4 meters.

Jhana worked with several classmates to measure and calculate the area of the school's hallways and used ratios to compare the hallway space per student in her school, Francis, to the hallway space per student at Longmore. She argued that ratios "make it easier to see the big difference" and noted:

> [Before] I wouldn't really use math. I would just say, LOOK how much space they have [in their school] instead of what we have [in our school]. ... But I would really use math now. ... Math made my argument make more sense, and have more of an idea, and actually tell what is happening, because it gave more detail to it.

As the class continued to analyze overcrowding at their school, they discovered disparities between their own space and that of other schools, and numerous instances where their school violated district building codes.

For example, during one lesson after students had worked in teams to measure and calculate the area of different classrooms and hallways, Beatriz asked each group to share their measurements. Students were shocked as they viewed the size of their own classrooms (e.g.,

474 square feet, 497 square feet, 567 square feet) compared to the classrooms at Longmore (e.g., 772 square feet, 864 square feet, 918 square feet). One student commented, "It's not fair! They have a smaller amount of students and bigger classrooms. They have to keep cutting our classrooms in half because we have so many kids."

Students' participation in this unit helped develop their senses of themselves as people who make a difference.

Ultimately, students decided to share this information with the district. They wrote letters to the superintendent, prepared fact sheets with the results of their analysis for administrators, and spoke at a school governing board meeting. As Jhana considered whether the class's analysis of the space crisis at their school made a difference, she commented: "Yes [we made a difference], because first of all, we found out something for ourselves, and we actually proved a point. ... WE made the difference. ... We learned what we learned, and we told people. ... Math made our argument make more sense. .... You couldn't do it without the math."

As mathematics educators, we would like all of our students to exhibit such passion about the power of mathematics. Jhana and her classmates invented novel problem-solving strategies and used mathematics as a tool to analyze and act upon situations at their school. We believe that students' participation in this unit helped develop their senses of themselves as people who make a difference.

Students Negotiate the Curriculum

Throughout the "Overcrowding at Our School" project, the students had opportunities to insert their interests, goals, and purposes into the curriculum. For example, after several days of measuring classrooms and calculating areas, students formed small groups to pose their own problem about a particular aspect of the school space. Beatriz asked students to identify one issue dealing with overcrowding at the school and to discuss how they might use mathematics to find out more about the situation. As students posed problems that mattered to them, their desire to understand and affect the overcrowding increased their engagement in mathematics, and thereby enhanced the learning that occurred.

Angel, a tall and rather quiet African-American student, was not a frequent participant in problem-solving discussions before this unit. But when the class began to investigate overcrowding at their school, there was a notable shift in Angel's level of engagement. Angel was extremely concerned about the school's bathrooms. She found it difficult to navigate among the other 10 or 12 people in the tight space. She noted that all females in the school, including 103 students and 15 teachers, had to share one rather small facility with only three working stalls, and a very small sink station. So when Beatriz asked Angel's group what aspect of the school space they wanted to investigate, the choice for Angel was obvious: "We want to know: Why are the girls' bathrooms so small?"

Angel's group constructed a floor plan of the restroom, measured its dimensions, calculated the area, and then analyzed the bathroom space based on the number of stalls, the estimated wait time during peak use periods, and the space available for waiting. Angel spoke about how the opportunity to investigate an issue she cared about made her feel "mad curious" and drew her into the mathematics. She commented:

> It was easier to do the math this way, instead of just learning it straight, like solving a problem, because we would actually, like, really get into it, and that made it easier. ... Like the facts [about the

school], they made you want to find out the answer. Like we wanted to know."

For other students, the opportunity to investigate real issues not only increased their engagement, but also pushed them to construct and apply important mathematical concepts.

Lianna, like Jhana, was concerned about the school's narrow and densely populated hallways. As she left Beatriz's classroom each day, she faced the continual challenge of navigating through one of the school's narrowest and most densely populated hallways. Not a student who was comfortable pushing her way through oncoming crowds of up to 80 children at a time, Lianna was often left standing just outside the door for four or five minutes while other students passed in and out of adjacent classrooms. Her group decided to compare the total hallway area at Francis with the area of the hallways of Longmore. The mathematics that Lianna's group engaged in would qualify as rigorous in any sixth-grade classroom. They developed their own strategies for multiplying mixed numbers to find the area of hallways with dimensions like 18-3/4 by 1-1/4 meters, and subdivided irregular spaces into rectangular and triangular areas.

But Lianna was not content with simply stating the total hallway area of each school; she wanted to make her argument stronger or, to use her words, to "use more specifics so people will listen." When she overheard that a classmate, Thomas, had calculated the ratio of hallway space per student, she was intrigued.

"How did you do that?" she asked. "We already found out the [hallway] area of Longmore, and I want to see how much [space] they will each get. You found out how much each person will get in Francis, and I want to do the same thing in Longmore. But I don't know how to do it."

"You've got to know how many students there are," said Thomas.

"Sixty," said Lianna.

"Sixty students. And how much is the area?" asked Thomas.

"Two hundred forty-six and 3/4 meters squared," she answered.

"So I am going to divide 60 into 246," said Thomas. "Because that way I can find out how much each person gets, cause it kind of divides it [the space] up.

Several days before this conversation, Beatriz had presented a mini-lesson designed to help students think about overcrowding in terms of "space-to-people" ratios. Many of the students found that comparing ratios helped support their claims of overcrowding, and some of them, like Thomas, had become adept at using this mathematical tool. It was not uncommon for students in the classroom to ask each other for help or to question each other's strategies in order to understand them better, as Lianna did. With Thomas's help, she figured out that if all students at Longmore entered the hallways at the same time, each student would have 4.1 square meters to her/himself. She was shocked when she compared that figure to the less than one square meter of hallway space per student at her school.

Taking Action

We wanted to support students in sharing what they learned through their investigations with the school and neighborhood community. Beatriz helped students to brainstorm ways that they might educate others about overcrowding at the school. Students generated lots of ideas, including distributing flyers, visiting the school board, going "on strike," making a large floor plan of the school to display, and compiling all their data to share with the district. Except for going on strike, the students implemented all of these ideas.

At the end of the unit, Naisha, a spirited and opinionated African-American student, spoke at a school advisory council meeting. This council represented the school at the district

> "The 3,000 participants at the [2002] World Economic Forum, which drifted through the hallways of the Waldorf Astoria, dropped $100 million on New York hotels, ballrooms, and restaurants, according to the New York City tourism board. That comes out to $33,333.33 per person. In five days in New York, each participant of the World Economic Forum spent on average what the average American makes in a year, four times what the average Mexican makes in a year, 14 times what the average person in India makes in a year, 22 times what the average person makes in Bangladesh, and 74 times what the average person makes in a year in Sierra Leone, according to United Nations figures."
>
> — **Derrick Z. Jackson, "The Elite's Pure Greed,"**
> *The Boston Globe*, Feb. 8, 2002

level and helped make decisions on matters of spending, curriculum and assessment, staffing, and enrollment. Naisha embraced the opportunity and volunteered to prepare a speech. What follows is the text of the speech she presented to the board:

> Good evening, my name is Naisha Watson. I am a sixth grader at Francis Middle School, and I am going to talk about overcrowding at our school. Our math class has been comparing our school to Longmore. We have noticed, as a class, that we have no space for kids to sit. ... The board of education has a building code that the classrooms have to be at least 750 square feet for 30 children. As you can see on the graph, only 3 classrooms are big enough, the rest of the

classrooms that are orange on the graph are smaller than 750 square feet. [Refers to a large diagram of the school created by several students.] ... The board of education has another building code that says the hallways must be 5 feet and 8 inches wide. ... There is only one hallway that is 5 feet and 8 inches. All the other hallways that are red on the graph do not meet the board of education building code. ... In our school we have 213 kids. If there was a fire in our school it would be a hazard to get through our narrow halls. So, as a school, we think we should have less students or more space.

Naisha felt that speaking out as a way of resisting the inequities her class discovered was not only necessary, but also potentially effective. "I think it's good [that we talked to the district], because if you keep talking to them then they will probably listen," she explained. "And you will get on their nerves and maybe then they will want to give us more space, or let us be in

The opportunity to investigate real issues pushed students to construct and apply important mathematical concepts.

a different building with more space, [space] that is lawful."

Even though the students knew that the district lacked funds to build a new school or add a floor to the building, they felt good about contributing to the public discussion of overcrowding at the school. As Lianna argued, "We have to say something because we are the students and we are the ones that have to live in the school everyday."

Challenging Students' Ideas

Not only did opportunities to engage in responsive action support students' sense of themselves as people who can and do make a difference, but using mathematics as a tool to support their actions challenged students' view of the discipline. For instance, when we initially asked Naisha what she thought of mathematics, she responded, "What do I think of math — you mean, numbers?" She described math as something she felt good about only when she got the answers right.

In contrast, as Naisha reflected on Beatriz's class at the end of the semester, she explained that unlike previous classes, where she studied material but never had the opportunity to "do anything" with what she learned, in this class, "we did something with it. ... Without the math, then, we wouldn't have the area of the school, and we wouldn't really know. And the [district] meeting wouldn't have been as powerful as it was."

Naisha was not the only student who began to recognize that math "made [her] arguments make more sense." Other students said math helped them to "prove how most stuff is not shared evenly," and "to prove to the district that our school was smaller;" math "gave more details" and "specifics" to their arguments and afforded them "more defense" in the problems they were fighting against. Students also spoke eloquently about how they drew upon mathematics to address "things in [the] community and school," and referred to this way of engaging with the discipline as "a lifelong thing" that is not only about mathematics but also about "things that you be in everyday, and it's a part of your life." Given that students often struggle to identify personal reasons why they should learn mathematics, these shifts in their understanding of the discipline are significant.

Further Reflections

As we reflected on the project, we found that creating space for students to pose their own problems and to inject their interests and concerns into the curriculum was a powerful way of supporting student activism. Occasionally, students posed problems about the school that did not lend themselves to rich mathematical investigations. We recognize that teachers have a responsibility to ensure that students learn certain content, and students' interests may not always lead them to a given mathematical idea. Beatriz had clear mathematical goals in mind for this unit (linear and area measurement, ratio, operations with fractions, and mixed numbers). But she thought it was equally important that students participated in mathematics projects that were personally and socially meaningful. We acknowledge that mathematics may not always be the best discipline to address the questions that students pose. Beatriz's challenge was to work with students to negotiate an intersection between their interests and the mathematical content they needed to study.

We also found that creating a classroom culture where critique was welcomed, and even expected, was essential. It was important for students to feel safe posing difficult questions, such as those that alluded to the connection between their particular situation at Francis (a poor school with inadequate space and resources), and broader educational inequalities that exist along the lines of race and class. While classroom discussions began with matter-of-fact statements about the data students were collecting and what they disliked about their school space, through questions such as, "Why do you think it is like that?" the talk gradually shifted to an exploration of why there were particular discrepancies, namely those between their own neighborhood school and Longmore. Students argued that the superior conditions at Longmore (a more affluent school) were not random but were directly related to the race and socioeconomic status of the students. "There are more white people than anyone else at that school," one student noted, and "the white people always get the good education, it's like an upper-class thing, for the white kids. ... That's just how it is!" In further discussion, students suggested that the demographics (in particular the socioeconomic status) of Longmore students were unlike those of Francis, and that these differences in demographics might be linked to the discrepancies in both the condition and size of school facilities, and to the "protection" of the school by those in power.

Beatriz opened a space in her classroom for students to approach their situation with a critical mindset, and in doing so, she supported their sense that they can act and make a difference.

At the end of the study, it was still unclear whether the district would increase the school's allocated space or make any adjustment to the number of incoming students that Francis had to accept. So unfortunately, the students ended sixth grade not knowing whether their efforts had any direct impact. But over the summer, the district decided to reduce Francis's incoming class by approximately 30 students, which allowed the school to retain its current size instead of increasing from 213 to 240 students, as initially planned. The district's action prevented an already overcrowded school from growing, which students welcomed as one small success.

Yet, what seems most important is not whether this particular "battle" was won or lost, but the shifts in understanding and increased critical awareness that students took from the experience. As Jhana put it: "We found out something for ourselves, and we actually proved a point. We made a difference. Math made our argument make more sense. You couldn't do it without the math." □

All students' names have been changed.

DESIGNING A WHEELCHAIR RAMP

PUTTING THE PYTHAGOREAN THEOREM TO USE

BY ADAM RENNER

As a geometry teacher and community-service-club moderator, I constantly sought more critical and authentic ways to teach secondary math. Taking a project-based focus by my third year, I also wanted to connect math to other disciplines and to issues of social justice.

One project involved the design of a community garden: Students designed scale drawings, estimated costs, and investigated issues of hunger and poverty in the local and global communities.

Another example involved designing a wheelchair ramp on campus. Since ours was an old, privately run school, we had somehow avoided mandates to make the building handicapped-accessible. While the building did have an elevator, the only ground-level entrance was a door inconveniently located in the basement, at the rear of the school. When it came time to teach the Pythagorean theorem during my second year, I decided to have the students design a wheelchair ramp. Students were required to figure out the best place to put a wheelchair ramp, develop a scale drawing, and contact a local contractor to find out the building cost based on their measurements and calculations. Measuring the height of the steps and knowing that for every one foot of rise, 12 feet of run were required, students would have to use the Pythagorean theorem to calculate the length of the ramp: the hypotenuse.

THE FOLLOW-UP

When students presented their designs and the contractors' bids for the wheelchair ramp to the school's administration, the students were thanked for their hard work, but their suggestions for building the ramp ultimately were ignored. Our follow-up discussion in class fruitfully centered on issues of public versus private facilities, the federal Americans with Disabilities Act, the fact that no students in this school were, at that time, disabled, and the apparent budgetary priorities of the school.

During my third year of teaching, when it came time for the Pythagorean theorem, students knew the administration previously had dismissed suggestions that a contractor build a wheelchair ramp. This time, the students offered to build the ramp themselves as a community-service-club project. Once again, the students were thanked for their work, but the plan was rejected. We discussed the same issues the students had covered during the previous year.

Sadly, later that year, a student was involved in a car accident which left her in a wheelchair. Watching her enter the school, day after day, via the rear of the school was a sad reminder of the denial of the students' work.

FIVE YEARS LATER

Five years later, after I had left the school, there was still no wheelchair ramp. As a more experienced teacher, I now wonder how I might have handled the situation differently. Should we have attempted to garner more support from others at the school? Should we have been more visible in our opposition to the administration's decision? Should we have sought legal help? I will always wonder if I missed an opportunity. ■

CHAPTER TEN

Radical Equations

BY DAVID LEVINE

Radical Equations: Math Literacy and Civil Rights. By Robert Moses and Charles E. Cobb Jr. (Boston: Beacon Press, 2001.) 233 pages. $21.00 hardcover.

In *Radical Equations: Math Literacy and Civil Rights,* veteran civil rights activist Robert Moses collaborates with journalist Charles E. Cobb to offer a stirring account of the Algebra Project, a reform initiative designed to help African-American students achieve a high level of mathematical competency. The book raises important issues about both math education and the struggle for racial equity within our schools.

The Algebra Project focuses mainly on the middle school years, when Moses and his colleagues believe African-American children must be prepared to enter high school math classes, which will open the door to higher education and technical careers requiring a strong math background. It encompasses new curricular materials, teacher training, the development of student leadership, and community involvement well beyond the scope of most educational reform efforts. From a modest beginning in Cambridge, Mass., the program has grown into a national network with 18 sites, over 100 schools, and 40,000 students.

As a member of the Student Nonviolent Coordinating Committee (SNCC), Moses pioneered voter registration work in Mississippi during the early 1960s. Through his soft-spoken courage and patient encouragement of local leadership, he played a crucial role in building the movement, which overturned state-sanctioned segregation and disenfranchisement in the South. After a sojourn in Tanzania, where he and his wife, Janet, taught school, he and his family moved to Boston.

By 1982, Moses had been tutoring his eldest child Maisha in math for years. He believed she was ready for algebra, a subject not offered for eighth graders at her Cambridge school. Since Maisha rebelled against having to do "two maths" — her regular schoolwork and the algebra tutorials her father insisted upon — he convinced her teacher to let him come to school to tutor her during the day. Soon he was working with a small group of students, and the Algebra Project was underway. As the program grew, it also became a family collaboration — Moses' wife and children, Maisha, Omo, Taba, and Malaika, all came to play important roles. In the early 1990s, Moses convinced a colleague from his Mississippi days, Dave Dennis, to bring the program into the Delta. This work has grown into a multi-state "Southern Initiative" of the project, which Dennis directs.

In their book, Moses and Cobb (a SNCC field secretary in Mississippi from 1962 to 1967 and now a senior writer for allAfrica.com) present the Algebra Project as a spiritual descendant and practical continuation of their organizing in Mississippi 40 years ago. They argue that the Civil Rights Movement's undeniable achievements in winning civic empowerment and formal equality for African Americans failed to overcome the economic servitude still endured by millions of black Americans.

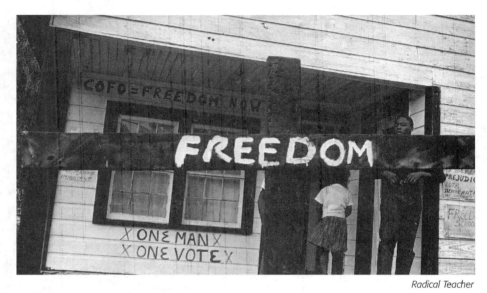

Radical Teacher

THE WORD "FREEDOM" APPEARS IN WHITE PAINT ON A CROSS BURNED OUTSIDE A FREEDOM SCHOOL IN PASCALOUGA, MISS.

This failure has been exacerbated by profound technological changes. Farm mechanization has reduced the 110,000 agricultural jobs in the Mississippi Delta during the 1960s to just 17,000 jobs today, reflecting a national erosion in semi-skilled and unskilled jobs in the industrial sector. At the same time, the computer revolution has generated the need for "knowledge workers" with strong academic skills. Cobb and Moses contend that poor (and poorly educated) white, black, and Latino students of today are the equivalent of Mississippi's disenfranchised black sharecroppers of the 1960s, "trapped at the bottom with prisons as their plantations." More specifically, they argue that mastery of the increasingly technological workplace depends on increasingly sophisticated math skills, including algebra. "People who don't have it [algebra] are like the people who couldn't read and write in the industrial age," they argue.

To help African-American students master mathematical literacy, the program has replaced traditional, rote-bound instruction with imaginative activities that engage student creativity and encourage sophisticated mathematical reasoning. An African drums curricular unit is designed to

The grassroots organizing philosophy of the Algebra Project offers a dramatic departure from many mainstream reform efforts.

pair a drummer and a teacher in lessons which teach fourth and fifth graders about ratios, proportions, fractions, and rates. In his work with high school geometry classes, Moses encourages students to post their own versions of geometric proofs on the classroom wall, to be analyzed and possibly challenged by their classmates.

For the sixth-grade curriculum, which forms a bridge from arithmetic into algebraic thinking, Moses designed a five-step learning process. The students first observe or experience a physical event. For example, in a unit on positive and negative numbers, Cambridge students begin with a subway ride during which the teacher asks questions that focus their attention on their shifting environment. They then draw pictures, construct models, or in some other way create a representation of the event. The following step is to write a description of the event in their own language. Next, each class member translates her or his description into "regimented English," highly compact language which moves them into a mathematical mode, and from which they finally render the event as a mathematical expression. This five-step process helps students

gain a firm grasp of mathematical ideas, connect math to everyday life, and become comfortable communicating in the language of mathematics. Similar classroom practices in geometry and algebra courses encourage students to debate mathematical problems and actively construct their own understanding of math concepts.

The Algebra Pro-ject's pedagogy is not unique. It resonates with the experiential, inquiry-based approach advocated by the National Council of Teachers of Mathematics (NCTM) and resembles intellectually robust math instruction that can be found in some classrooms around the country. But the grassroots organizing philosophy of the program offers a dramatic departure from many mainstream reform efforts.

Moses believes that math education innovations are often implemented by university researchers whose primary frame of reference is their own discipline and academic community, and the modus operandi is to offer pre-packaged programs to schools. In contrast, the Algebra Project works on the premise that oppressed people can only win just schools through political organizing. To emphasize this perspective, the early part of the book describes how Moses and other civil rights activists built the Mississippi movement during the 1960s. With the guidance of Ella Baker, an experienced veteran of the black freedom struggle, Moses and his companions learned to develop the capacity of "ordinary people" to act as leaders and collaborate to bring about fundamental social change. Their approach, with its patient emphasis on democracy and nurturing the talents of poor people, has come to be known among civil rights historians as the "organizing tradition" of the movement. It is often contrasted with the "mobilizing tradition" of Martin Luther

King Jr. and other charismatic leaders, which is successful at turning out large numbers at demonstrations but often neglects the day-to-day work that builds powerful and sustained grass-roots involvement.

For the Algebra Project, "organizing in the spirit of Ella" rests on three principles:

1. The centrality of families to the work of organizing. When Moses and other young organizers reached the Mississippi Delta, they connected with strong local leaders. Often, these leaders would involve family members in the movement, helping to create crucial networks of political activists. The Algebra Project seeks to involve the families of students and other community members in committees which run the local projects.

2. Organizing in the context of the community in which one lives and works. The young civil rights workers were absorbed into local families, who fed and housed them and protected them from hostile whites. This helped the activists "sink deep roots into the community." The Algebra Project also operates on the idea that staff members should be fully immersed in the communities which host local projects.

3. Young people need to be empowered to fight for their own liberation. Moses points out that high school- and college-age young people provided some of the crucial leadership of the Civil Rights Movement. He believes that the reforms necessary for young black people to achieve deep math literacy will only come about when they become ardent and savvy advocates for their own education. Through the program's Young People's Project, for example, students tutor their peers, lead workshops for students and adults, and help plan and run math youth camps during the summer.

Algebra: A Key to Economic Liberation?

One of the book's core arguments is that students must master algebra to succeed in the workplace of the future. They cite Labor Department statistics that 70 percent of current jobs require "technology literacy" and that by 2010 all jobs will require "significant technical skills." Increasingly, essential technological expertise has come to mean relatively sophisticated understanding of how to use computers to perform a multitude of vocational tasks. To fully master computers, they argue, students need to be comfortable manipulating symbolic representations which represent "underlying mathematical concepts." They further argue that our society has designated algebra as the place where young people acquire such skills.

This cornerstone argument needs further documentation to be fully creditable. The phrases "technology literacy" and "significant technical skills" are quite general. We need to know if such literacy and skills specifically include algebraic thinking. Another issue is the varied impact of increased computer use on different occupations. The computerization of a job does not always bring the need for more sophisticated intellectual skills. Many low-paying service jobs have incorporated computer use which requires learning some new procedures, but not mastering substantially more demanding cognitive tasks. The authors would have been more persuasive if they had offered concrete examples of how algebraic skills are used in particular jobs, and evidence that such jobs are or will become a major part of our evolving economy.

Nevertheless, Cobb and Moses are not wrong to assert that algebra functions as a crucial gatekeeper to full economic opportunity. Even if a young person is not drawn toward a highly technical vocation, high school algebra is usually required for college entry. In addition, algebra provides knowledge necessary for advanced math which prepares students for a number of technical and scientific careers. Too many students of color lose these options through poor math performance before they reach high school. As Cobb and Moses note, part of this problem is reflected in Ph.D. statis-

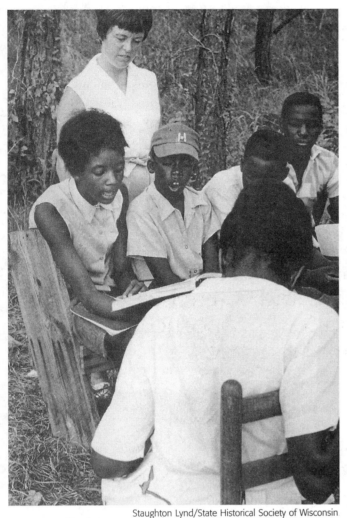

Staughton Lynd/State Historical Society of Wisconsin
A CLASS AT A MISSISSIPPI FREEDOM SCHOOL IN 1964.

tics for technical fields. In 1995, blacks were 15 percent of the U.S. population but earned "only 1.8 percent of the Ph.D.s in computer science, 2.1 percent of those in engineering, 1.5 percent in the physical sciences, and 0.6 percent in mathematics." Finally, even though the authors could have presented stronger evidence regarding the relevance of algebra to adult employment, the technological evolution of many occupations does support their case. An understanding of algebraic concepts can help workers become more adept at working with spreadsheets, graphs, and databases. Our computer-based economy increasingly calls for such skills, even outside of highly technical fields.

Is the Program Working?

In assessing initiatives such as Algebra Project, a crucial question is whether the program is meeting its stated goals.

In Bessemer, Ala., teachers at Hart Elementary, a school of mostly poor, black children, started participating in the Algebra Project in the fall of 1991 while teachers at the predominantly white West Hills, one of the "top elementary schools" in the district, continued with traditional math instruction. During a three-year study initiated in 1995, Hart moved from trailing West Hills on standardized math tests by several points to exceeding it by a few points, compiling the highest scores in the district.

Radical Equations and other Algebra Project reports are filled with similar success stories. They also document instances in which Algebra Project students register in greater numbers than their peers in higher-level math courses. As Cobb and Moses tell the Algebra Project story, they weave into their narrative extended testimonials from parents, teachers, and students which provide both penetrating explanations of the reform process and many examples of how the program has helped students learn more.

While the vignettes and overall narrative thread give us a persuasive picture of an effective reform movement, the book would have been strengthened by more systematic documentation and analysis of the program's impact on student achievement. We need to learn more about the extent of the program's success in strengthening students' math abilities, and the classroom dynamics which make such success possible. In-depth case histories of Algebra Project classrooms would be helpful, as would comparisons between the learning experience of students within the program and the learning experience of similar students in traditional math classes.

Research on the project should not fall victim to the popular and crude trend in

American education to judge programs mostly by narrow quantitative measures. Cobb and Moses cite increased standardized test scores to document the program's success but realize that such numbers tell only a small part of the story. They examine the Algebra Project's impact on student motivation and work habits, teacher attitudes and behaviors, and community involvement. Future research should build upon and extend this holistic approach.

Algebra Project Lessons

In a review of school reform during the past century, educational historians Larry Cuban and David Tyack note that innovations often falter because their advocates fail to win political support. *Radical Equations* does a good job of teasing out insights from the kind of political work which builds durable support for substantive changes in how schools function. It offers a refreshing contrast to glib and self-congratulatory recipes for fixing up schools.

Even after nearly two decades of nurturing the program, Moses writes, "I have thought of the Algebra Project as a young child who is trying to stand up and teetering and falling down a little, then getting back up." The book pays careful attention to this teetering up and down of small groups of people trying to make their schools better. Cobb and Moses glean insights into the often contentious dynamics of school change from battles with the constraints of rigid standardized testing, uneasy administrators, and bureaucratic fear of innovation. The challenges faced by the Algebra Project affirm what they learned in Mississippi: People have to be willing to change themselves if they are to develop the strengths they will need to change the system.

Beyond the issue of math instruction, the Algebra Project offers compelling lessons on how determined networks of educators, parents,

and students can build a program which advances educational equity. Such democratic renewal promises the obvious rewards of promoting academic and vocational success for young people. But perhaps just as important, it also affirms local people's cultural values and capacity to deepen community life through shaping the public institution most likely to have a profound impact on their children. "Organizing in the spirit of Ella" means school reform which enriches the lives of teachers, community members, and students.

An African drums unit pairs a drummer and a teacher in lessons about ratios, proportions, fractions, and rates.

In contrast to top-down reform initiatives which demean the expertise and professional pride of teachers, Moses and his colleagues have developed training programs which build upon their strengths. A Cambridge teacher comments, "Bob was affirming what we were doing while he was helping us change. He didn't come in and say, 'We're throwing this out, it's junk.' He came in and said, 'You guys are great. Wanna try something different?' When we asked, 'How will it work?' he turned around and asked, 'Well, how do you think it should work? What do you want to have happen?'" By posing problems rather than solutions, Moses invites teachers to confront and work through the frustration and anxiety of experimenting with new ways of teaching.

Such collaborative processes within the classroom are buttressed by efforts to involve community members. Although the dynamics of community involvement differ from site to site, the project is deeply committed to encouraging local control. During a 1998 visit to the Jackson, Miss., Algebra Project, I talked with Kathy Sykes, who served as a project staff member and representative on the local Site Planning Committee. This group reviewed the program budget, helped plan such activities as student retreats, and encouraged parents to serve as chaperones for program activities. The committee also encouraged parents to sit in on classes and eventually hoped to train parents as classroom assistants. Sykes told me, "I feel this is sort of like a crusade. ... I think that the work which is going on here will make a difference in the lives of our people and that's why I want to do what I can to see that it continues."

The program seeks to instill this spirit of personal responsibility through pedagogy which encourages students to break out of their own passivity and take charge of their own learning. Mary Lou Mehring recounts how 12-year-old student Andrea Harvey asserted, "I'm going to do four lessons a week because I want to finish such-and-such by the seventh grade, so that I can finish the book by the eighth grade, so I can be in honors geometry in the ninth grade." Andrea went on to work with the Algebra Project and eventually become certified to teach math in the Boston schools.

As a continuation of the Civil Rights Movement, the Algebra Project places itself firmly in the tradition of education aimed at racial equality. At the same time, Moses conceptualizes the goal of the endeavor almost exclusively as improved job opportunities. The program does not appear to directly use math instruction to help young people see full citizenship as the opportunity to use their math skills to promote social justice. As many articles in this book indicate, math can be used to analyze social inequities within our society — such topics as the disparities between rich and poor school districts, the mathematics of sweatshop economics and the quantitative injustices built into the wealth and income structure of our society. Such themes might represent a fruitful direction as the program's curriculum evolves.

However, the absence of political math content hardly means the program is apolitical.

The authors persuasively echo Ella Baker's assertion that demanding something which is essential to your life and which you are systematically deprived of is an inherently radical act. Moses approvingly cites instances when young people agitate to make their schools dramatically improve math instruction.

For African Americans, the struggle for education has always been entwined within the struggle for freedom. This intimate historical relationship is underscored when the authors quote Mississippi school-desegregation activist Mae Bertha Carter: "The way to control black people or anybody is to keep them dumb. Back in slave time they catch you reading and they would whip you. Education, that's the goal. These [present day] school systems ain't doing nothing but handicapping these children."

In a society so afflicted with faulty historical memory, the Algebra Project demonstrates the necessity of learning from our past to fashion our future. In doing so, it puts history to its most honorable and practical use. □

FOR MORE INFORMATION ON THE ALGEBRA PROJECT

Algebra Project Inc.
99 Bishop Richard Allen Drive
Cambridge, MA 02139
617-491-0200
www.algebra.org

Young Peoples' Project Inc.
99 Bishop Richard Allen Drive
Cambridge, MA 02139
617-354-8991
www.typp.org

CHAPTER ELEVEN

The Geometry of Inequality

BY ANDREW BRANTLINGER

taught a summer course several years ago to a group of students in a neighborhood school on Chicago's north side. They were there because they had failed geometry during the regular school year. Most of the students were non-white, and 85 percent of the school's students were classified as low income. Many of the students in my class were not on track to graduate in four years and had a history of failing mathematics.

When I interviewed my students, many of them said they didn't see how mathematics related to their present and future lives outside of schools. So I chose to teach a lesson I hoped would help them connect geometry to issues of fairness and see how math could be relevant.

I began the lesson by asking my students if they had heard of Rodney King and the riots that took place in South Central Los Angeles in 1992. The five African-American students claimed that they knew who Rodney King was, while many of the other students did not. One of my female African-American students explained that Rodney King "got beat by four white cops" and described a bit of what she knew about South Central. I added that the disturbances broke out in South Central and other neighborhoods in Los Angeles after a mostly white jury acquitted these four white police officers.

After this brief introduction, I asked the students to work in groups to come up with what they felt to be average ratios of people to movie theaters, liquor stores, and community centers. Obviously, a community's resources comprise more than simply theaters and liquor stores, and might also include parks, quality schools, libraries, diverse shopping areas, and adequate public services like water and sewage, not to mention access to good jobs. But the figures I had came from a National Public Radio story that came out shortly after the riots in 1992, and I was really struck by the disparity in South Central between the number of theaters and community centers on the one hand and the number of liquor stores on the other. I decided to use these statistics as the core of my lesson.

Initially, three of the four groups were reluctant to come up with estimates for the three people-to-resource ratios. I think this was, in part, because I was asking them an unusual school mathematics question — one with no exact answer. A few students insisted that I had a "right" answer in mind, despite my reassurances to the contrary. In some sense they were correct; I *did* push them to justify estimates that seemed unreasonable to me.

Movie Theaters and Liquor Stores

Next I asked the groups to use their people-to-resource estimates to make predictions about the number of movie theaters, community centers, and liquor stores in a "three-mile radius" in South Central circa 1992. Having used this activity with preservice teachers a few months earlier, I realized that estimates for the

number of community places in South Central can be made without doing much mathematical thinking. I required my students to engage with the geometric issues that the problem presents. I hoped a better geometric understanding would improve students' critical understanding of the situation.

Because I also wanted my students to connect their everyday knowledge — of things (population distribution and size of city blocks) — to the problem, I presented each of the four groups with a map of Chicago and asked them to draw a circular region with a three-mile radius on it. Again, I wanted students to think mathematically, so I asked the class to figure out how many square miles and square city blocks would be in such a region.

Some students responded by drawing a square-mile grid on the map, while others used more ad hoc approaches to finding the area. As I went from group to group I heard them ask each other: How big is a square mile? Where is a square mile on the map? Can we use $A = \pi r2$ to find the area or do we have to count squares? When two of the groups asked me if $A = \pi r2$ would give them the same result as counting the number of square miles inside the circle, I responded, "Try to find it both ways and see if they give you the same answer."

Approximating the number of city blocks in a circular region with a three-mile radius fostered a good discussion, in part because there was some degree of confusion. That is, my students were used to thinking about linear blocks as a measure of distance but they were apparently not used to thinking of square blocks as a measure of area. One group of students asked me if they could find the area of a square mile by counting the linear blocks on the perimeter of a square.

A student in a second group initially argued that the area of South Central would be 48 blocks because the diameter of the circular region is six miles and there are eight city blocks per mile. But one of her groupmates disagreed, stating, "No, it should be 96 blocks because you have to fill the circle up on both sides of the diameter." I asked this second young woman to show us the 96 blocks on the map. Shortly after beginning her count, she looked up, smiled, and claimed that 96 blocks would be far too few to fill up the circular region.

When I returned to this group later on, they explained that they used $\pi(24)2$ to calculate the approximate number of blocks (1,810) in the circular region. This solution — using a radius of 24 blocks instead of three miles — was slick. I had not considered it myself. When I asked them if they were sure of their answer, they showed me that they could fill up a quarter of the circle with rectangles, which when combined would give an area of about 450 blocks, one fourth of 1,800.

Working through the geometric aspects of the activity took the better part of a class period. At the end of the day, I asked each student to jot down a new estimate of how many liquor stores, community centers, and movie theaters they thought were in South Central circa 1992. I was a bit frustrated that several students were reluctant to give estimates. I had assumed that this mathematical work would make answering this question easier. Four or five of my students insisted they would first have to count movie theaters, community centers, and liquor stores in their own neighborhoods in order to come up with an answer they felt comfortable with.

We didn't have time for students to count in their neighborhood, so I provided them with data from Evanston, a mixed-income suburb that borders Chicago to the north. When we revisited the South Central problem the next day, I put an overhead up that stated that Evanston was a "fairly typical community" with an area of 8.5 square miles with approximately 75,000 people,

> # I wanted my students to connect their everyday knowledge of things to the problem at hand.

seven liquor stores, eight community centers, and three movie theaters. (I used the Evanston yellow pages to estimate the number.) The students were able to quickly figure out that approximately three Evanstons would fit into the area of South Central and that a little more than three Evanstons would have about nine or 10 movie theaters, 26 community centers, and 27 liquor stores.

As my students read these results to me I told them, "I see this activity as a critique of our society as a whole, and our government, and not of the people who live in South Central." I said, "According to National Public Radio, at the time of the riots, there were zero community centers and zero movie theaters." This surprised them.

Then I asked the class to guess how many liquor stores they thought there were in South Central when the L.A. riots took place. They called out a range of numbers from 5 to 100.

When I stated that the actual number of liquor stores was 640, Dele let out an "Ooooh..." and Maya exclaimed, practically shouting, "What!?"

José said, "Oh man, that's cool," and laughed.

Dele said, "All they want them to do is drink." Then Dele added, "All they do is drink."

Upon hearing this second comment, I repeated that I did not intend for this activity to be a critique of the people who lived in South Central. "Look, there are no jobs," I claimed.

"That's why they be on the streets," responded Dele.

Later on, when students were writing up their individual responses to the activity, Dele's groupmate, Tony, asked me how I thought the overabundance of liquor stores in South Central was connected to the riots. I explained that I meant there was probably good reason to rise up, considering the lack of alternatives and the gross resource inequities between what people in South Central had compared to, say, people in Evanston. Tony looked me in the eyes and nodded, but I don't remember him responding to me verbally.

Potential Pitfalls

When I reflect on my "South Central" lesson I see two pedagogical problems. First, I had not expected this lesson to potentially reinforce the dominant worldview that the problem with South Central is the people who live there. My students know that some white teachers buy into prevalent myths that blame working-class blacks, Latinos, and whites for being poor and other social ills. So, in hindsight, it makes a lot of sense that Tony asked me to re-explain where I stood on the root causes of the riots.

The next time I do this activity, I will bring in additional resources in order to clearly set up the sociopolitical context of the problem and to hopefully avoid the possibility of reinforcing negative stereotypes. Excerpts from an article by Mike Davis called "Who Killed Los Angeles?" that discusses the root causes of the riots — i.e., economic disinvestment, racist government policy — would be helpful in this regard.

Second, I shied away from opening up the political whole-class conversation that might have happened at the end of the lesson. The students reacted to the shocking numbers, but then I spoke *at* them about my beliefs. While I did make an attempt to open up the floor, I should have asked them to discuss what *they* thought were the root causes of the riots and what they thought about Dele's two conflicting comments that "All they want them to do is drink," and "All they do is drink." I also wish I had asked Dele who the "they" was he was referring to.

Instead of pushing my students to enter the conversation, I rushed on to the planned end of the activity in which I explained Brazilian educator Paulo Freire's notion of "reading and writ-

ing the world." And I ended the activity by asking the class to write individual responses to the (perhaps leading) questions: "Is this activity a meaningful way to begin to read the world using mathematics?" and "Does the use of mathematics help one understand more deeply aspects of our reality?"

In truth, the problem of opening up a political conversation was not entirely due to my lack of pedagogical expertise; three of my students made it clear to varying degrees that they expected me to teach the depoliticized form of mathematics they were used to. One of my African-American students repeatedly opted out of critical activities like this one, choosing instead to work alone on geometry worksheets. I wondered if her actions meant she doubted my assertion that math could be socially relevant.

Despite its flaws, I believe this activity was powerful for many students. Many students did seem to think that we should discuss issues of fairness in school. Typical written student responses to the activity read, "We should have more jobs and more community centers, then you wouldn't have to worry about the riots," and "The government should put more money in [South Central]."

Finally, I think that Dele spoke for many of his fellow students when he told me they were learning "more about real facts" in critical activities like South Central than in decontextualized geometry activities. This intrigues me because I have heard many mathematics educators claim that traditional mathematics is as close to the truth or "real facts" as students will ever get. Clearly, Dele did not see it this way. □

All students' names have been changed. The South Central activity was adapted from an activity developed by Eric Gutstein. See p. 101.

SOUTH CENTRAL LOS ANGELES

RATIOS AND DENSITY IN URBAN AREAS

BY ERIC GUTSTEIN

This is a very open-ended activity, mathematically, and students may be disoriented because of that, so be aware of the possibility and adjust as necessary. It is intended to be given in parts, as indicated. It is important to finish the activity by having students share what they've written in part 3(b) with the whole class and discuss the different views. See www.rethinkingschools.org/math for background reading on the Rodney King beating and verdict.

This can be done in grades seven to 12; in the later grades, depending on the mathematics being studied, you can alter part 3(a) to make it more challenging. Resources like maps, almanacs with populations, and measuring tools can be used. Make sure students explain carefully their mathematical strategies, because there are no "correct" answers for these problems and many possible solution methods.

PART 1

a)   Three types of places that exist in communities in the United States are movie theaters, community centers, and liquor stores. What seems (to you) to be a reasonable ratio of each of these to people; that is, how many people to each movie theater, etc.? What are you basing your estimate on?

Note: People may differ about the definition of "community center." One suggestion is not to count churches, but things like fieldhouses in city parks, Boys and Girls Clubs, YMCAs, etc. Also, for liquor stores, ask students to think mainly about stores that really depend on alcohol sales, unlike supermarkets that also sell wine and beer (and even liquor). But these definitions are open for negotiation. The main point is for students to state explicitly any assumptions.

b)   Think about an "average" city in the United States — what does that mean to you? Now think about a city in the United States that is "average" in that it is relatively crowded (dense) but has mainly homes with some apartment buildings. In this "average" city, given your answers to part 1(a) above, what do you think would be a reasonable number of community centers, movie theaters, and liquor stores? What are you basing your estimate on?

PART 2

The year 2002 marked the 10-year anniversary of the "Rodney King verdict" in the California community of South Central Los Angeles. When the verdict was

101

announced, and the South Central community rebelled/revolted/rioted (depending on your perspective), National Public Radio reported on the actual number of movie theaters, community centers, and liquor stores in a "three-mile radius" in South Central. What would you estimate the number of each to be?

Note: Students may draw on background knowledge and/or media stereotypes to answer this. The point is, even if they do, try to get them to use mathematics to figure out how many people live in a three-mile radius in South Central, and then, using their answer to part 1(b) — because the community description in question 2 fits South Central — how many of each there should be, based on their work. Maps of students' own cities and data can be used to help figure this out. For example, students can figure out how many people live in such an area in Chicago if they are given Chicago's size (use maps) and population (about 2.9 million). Students can then adjust to South Central, which is somewhat less dense than Chicago.

PART 3

a)   As reported on the radio, there were zero movie theaters, zero community centers, and 640 liquor stores within the three-mile radius. What would you say is the density of liquor stores in the area?

Note: Density, in this case, means how many liquor stores per person and/or per city block. In South Central, 16 blocks is about a mile, e.g., from Florence Avenue (72nd Street) to 64th Street is one half mile. A good question to ask is: If the liquor stores are randomly distributed throughout the three-mile radius area, about how far does anyone need to walk from their house to get to a liquor store?

QUESTIONS

b)   Finish by asking students to write about what they have learned. Some questions you could pose to students are:

Why do you think these data are the way they are?

What mathematics did you use? What math did you learn in this activity?

How was the mathematics important for you to understand the social and political issues?

If this was your neighborhood, would you do anything about this situation? Why or why not? If so, what would you do? ■

CHAPTER TWELVE

Integrals and Equity

A Math Lesson Prompts New Awareness for Prep School Students — and Their Teacher

BY MEGAN STAPLES

An AP calculus class at a prestigious boarding school does not seem a likely venue for student reflection on privilege and wealth. But when I taught a group of academically inclined math enthusiasts several years ago, I made some important discoveries, and I believe I helped my students view their world with a wider lens.

It was my sixth year of teaching math and I loved it — the intellectual challenge, the complex puzzle that would suddenly sort itself when that key piece was put in place, and the interconnectedness that appeared among seemingly disparate ideas.

I took a new job at a small private boarding school in central Massachusetts. I was impressed with the way the school guided its 300 students to become moral, thoughtful adults. There was a schoolwide work program, so everyone pitched in to maintain the buildings and grounds. They had an extensive community-service program, where students volunteered at soup kitchens and nursing homes and periodically undertook larger projects such as building homes for Habitat for Humanity. The academics, of course, were rigorous and nearly 10 percent of that year's graduating class would find themselves with Harvard admission letters. The students were privileged, and the school knew it was preparing future leaders.

My AP class had a weekly "lab period," which provided a luxurious 90 minutes of class time compared to the standard 42 minutes. The class had this extra time to allow for AP exam test prep, but it also afforded the opportunity to conduct some more exploratory lessons, which I tried to do as often as possible.

As with my other lessons, I planned this one with objectives that were decidedly mathematical. We had been working with integrals, finding areas under various curves and calculating infinite sums, and I wanted to introduce my students to a real-world application of this mathematical concept by examining the GINI Index.

The GINI index is a measure of income distribution (or dispersion) across a population — in this instance, we considered households in the United States. In a single number ranging from 0 to 1, the index provides an indication of the equitability of this distribution. A GINI index of 0 corresponds to perfect economic equity — every household has the same amount of income. A GINI index of 1 corresponds to extreme inequity, where one household has all (100 percent) of the income and all others have none. These are theoretical extremes; in reality, the number lies somewhere in between.

The index is calculated by first dividing the population into income groups — for example, by quintiles (fifths) or percentiles (hundredths) — and finding the percent of total income "owned" by that group of households. The group that represents the lowest quintile, for example, will hold less than a fifth of the total income (the accumulation of all households). The top quintile will have more than 20 percent of the total income.

At right is one graph of income distribution from 1996. The accumulated percentage of income is plotted on the vertical axis against the percent of households (horizontal axis). The point (20,4) means that the lowest 20 percent of households held 4 percent of the total income. The point (40,13) means that the lowest 40 percent (the two lowest quintiles together) of households held 13 percent of the total income.

The GINI index is twice the area between the 45 degree line (representing perfect equality) and the income curve. Thus it measures the "gap" between perfect equality and the state of affairs described by the income curve. Note that the curve always includes the points (0,0) to (1,1). Its exact shape is what changes, depending on the distribution of income.

Estimating Income

After a brief introduction to the idea of income distribution, we began talking about income in the United States. Perhaps not surprisingly, my students made reasonable estimates when I asked them to guess what percent of total income the top 20 percent of the population held. And they made somewhat reasonable estimates for the other quintiles. They didn't seem troubled by the fact that the top 20 percent of households held 47 percent of the wealth. (I assumed they attributed the disparity to super-rich individuals like Bill Gates and Michael Jordan.) But the most meaningful conversation arose when I asked them to consider the median income of the various quintiles. The median represents the middle value of the set of incomes in the quintile, often close to the average. I was perplexed by their lack of reaction to the percentage

INTEGRALS AND EQUITY

1996 INCOME DISTRIBUTION

distribution, and I was curious what they thought the median incomes might be.

"Median income of the lowest quintile — what do you think?" I asked.

"Eighteen thousand," someone guessed.

"Twenty-two thousand," another said.

I recorded their estimates on the board. I didn't ask for any justification, leaving their reasoning private. Looking back, this is somewhat surprising as I was always asking my students "Why?" and "How do you know?"

I then asked them to guess the median income of the top quintile.

"One-hundred seventy-five thousand."

"Yeah, around there."

"One sixty." "Two hundred."

And others offered thoughts.

"And how about the next quintile down — this 60 to 80 percent group?" And we went through the rest. Once we had their set of values on the board, it was my turn to tell.

"Median income of the lowest quintile is $8,600," I reported.

The response was "Whoa," followed by some silence.

"That means half the households are below that — 10 percent of the people." Although this fact seemed to be mind-blowing for many, they did not challenge it.

I continued. "The next quintile up is $21,097. The third quintile is $35,486. The fourth quintile is actually $54,922, and the top quintile is $115,514."

My students expressed disbelief. "No way!" "Is that right?" I think the median incomes of the fourth and fifth quintile surprised them the most: They had thought that the fourth quintile would be around $90,000 and the top quintile would have a median income closer to $170,000.

Given their privileged backgrounds, perhaps I should not have been shocked that the students' conjectures were so far off. And perhaps I should not have been taken aback at their disbelief upon learning the median incomes for each quintile. Though I don't know the actual incomes of these students' families, this group was firmly situated in the top quintile, and most the top decile. These data were news to them — top headlines.

But I was surprised, perhaps most surprised, by the fact that the students had so little sense of their economic positions relative to the rest of society — this was despite all their community service, attention to issues of race and class in their history and literature classes, and interactions with classmates who were there on scholarships. Even with these experiences, the students inhabited a world where they could think of their families as more or less "average." In their social context there was very little evidence that would have led them to believe that only one in 10 people was as economically advantaged as they were. I don't know that this group of students had ever been asked to look at the "cold, hard numbers" before and pinpoint their positions relative to others. They knew they were from privileged backgrounds and had "more," but this quantification of how much more they had — orders of magnitude — was quite a surprise.

Worlds Apart: Math and Social Justice

The experience of seeing these students gain awareness of their positions in society helped me along in my evolution as a teacher. I had gone into teaching because I loved math

Math took place in my classroom. It existed a world apart from what took place in the dorm.

and loved supporting students in their journeys to become good people living in a challenging world. I also wanted students to be attentive to their communities and others around them, to understand that the world does not offer all an equal playing field. Ideally, I hoped they would use these understandings to change the world in ways that helped address some of the gross disparities that exist. But these were separate aspects of my "teaching." Math took place in my classroom. It existed a world apart from what took place in the dorm, on the soccer field or basketball court, during the van ride to the soup kitchen, or over an ice cream at the favorite local shop.

In retrospect, it's odd that these were separate worlds for me. I had been a math major at a progressive and politically active university. I had participated extensively in volunteer activities and lived in the Public Service House, a dormitory whose members were committed to community service. But math and being a good person in the world were separate for me there as well. My aspirations for social justice had not had a place in the math classrooms at the uni-

versity, and I didn't bring them with me to the mathematical part of my teaching. These worlds were not in opposition, but I didn't see them as natural bedfellows. Math was about the intellect and individual power. Being a teacher for me was about giving to society and connecting with individuals in ways that might make transformation possible.

It became clear to me that there was a large gap here, for me and my students. Despite my prep school students' general awareness and participation in activities, they remained unaware of their positions in that landscape. The hard and fast numbers brought this into relief.

I would like to say that I followed up on this initial lesson right away and implored students to examine other data that helped them understand themselves anew. We could have pursued the difference between men's and women's incomes or the incomes of different racial groups. We could have compared the United States to other countries. We could have looked at historical trends in the GINI index, and analyzed the factors that contribute to the greatest and least equity in income distribution over time. Regrettably, I did not pursue any of these.

Nonetheless, I was left with a powerful lesson. Through this activity I realized that those parallel pursuits could serve one another— where mathematical learning goals, even on a high level such as AP calculus, could be pursued at the same time as goals aimed at personal growth and social awareness. □

HIV/AIDS STATUS

USING HIV/AIDS STATISTICS TO UNDERSTAND RATIOS

BY SAGE FORBES-GRAY

Teacher Aim: Students will be able to interpret a statistic about HIV/AIDS transmission using their understanding of identifying, converting, and comparing ratios.

Student Aim: How do we understand HIV/AIDS statistics using ratios?

PROBLEM OF THE DAY

a) 1 in 250   b) 1 in 500

What are two different ways of expressing each of the ratios (a) and (b)? Which ratio is greater? How much greater?

The whole class will go over the problem of the day, focusing on equivalent ratios, coverting ratios, and comparing ratios. Possible answers for (a) will include 2 in 500, 1/250, and perhaps even 0.004 or 0.4 percent, and for (b) 2 in 1000, 1/500 and perhaps even 0.002 and 0.2 percent.

Hopefully students will see that 1 in 250 is greater than 1 in 500, and ideally students will see that 1 in 250 is twice as big as 1 in 500. Conclude by asking students: Where do they think these numbers come from?

MINI-LESSON

Students will be told that these ratios have to do with HIV/AIDS transmission. Be sure they understand the language we will be using about HIV/AIDS. If there is time, we will brainstorm briefly about what we all know about HIV/AIDS. We will use a "set-up" that lists truth, questions, and myths (one column for each).

Ratios can be used to describe statistics and are used to describe statistics involving HIV/AIDS. The aim will now be elicited.

The following slogan was plastered on the sides of New York City buses as part of a public health campaign called "Know Your Status":

"1 in 250 people in the United States have HIV, 1 in 500 know it."

I tell the students that we will be answering a few questions about a statement that I am about to put on the board. I will ask the students to copy the following questions and tell them that they will be working in pairs (or groups of four, depending on the maturity of the particular class).

QUESTIONS

1.   What does this statement mean?

2.   How could we restate this statement in three different ways? Try to use your understanding of equivalent ratios and converses to write these statements.

3.   Why do you think many people find this statement alarming?

4.   What can be done to change how people think and act about HIV/AIDS that would make this statement untrue, for better or for worse?

5.   What kind of statement would be a less scary and more hopeful statement? (You do not need to use the same numbers for this statement.)

6.   Do you think that this statement, which appears on the sides of many New York City buses, is a good way to get people thinking about the dangers of HIV/AIDS and the importance of knowing your HIV status? If not, how could the statement be improved so that more people understand and

are thinking about HIV/AIDS transmission?

As a class, we would review each question, focusing on any difficult language and reviewing equivalence and converses.

CLASSWORK

The students will be put in their groups and given 15 minutes to discuss the questions and then 15 minutes to write up their responses. I will tell the students that at the end of class we will take 15 minutes to discuss what each group thinks about the activity and that they should be prepared to present for two minutes and answer questions. The presentations will be informal: Students will simply share their answers to one or two of the six questions.

Next, I would reveal the statistics, put the statement from the bus ads on the board, and begin to move about the room answering questions and making sure that the students are on-task. I would remind them about their classwork grade and presentation at the end of class.

FINAL PRESENTATIONS

I would encourage students to make connections between this material in their own lives. Finally, I would explain the homework assignment, offering a sample and answering any questions. I would collect the work as they get ready to move on to their next class.

HOMEWORK

The students will be asked to write a letter to the campaign manager who designed the subway poster that they discussed in class today. In their letter, they would answer question 6 from the classwork and give their opinion to the campaign manager about how the subway poster affected them personally and how they think kids their age would be affected by the poster. The letters will be mailed to the campaign as a way to give the campaign manager real feedback from NYC youth. ∎

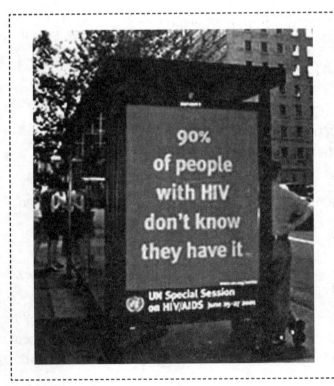

HIV STATUS

BUS-STOP PHOTO
TEACHING SUGGESTIONS

Have students research the number of people who are aware that they are HIV-positive. Then have them use the statistic on this bus-stop poster from the U.N. to calculate how many of those people don't know their HIV status.

Ask students to compare the statement in this photo with the ratio from the advertisement quoted in the activity above. Have them discuss why there might be a difference in the two statements.

JUSTICE FOR JANITORS

RICH LESSONS IN THE POWER OF MATH

BY BOB PETERSON

The picture book *¡Sí Se Puede!/Yes, We Can!* by Diana Cohn is a great picture book to use with students to show that math matters. The bilingual English/Spanish book, beautifully illustrated by Francisco Delgado, tells the story of a janitor strike in Los Angeles in 2000, when nearly 8,500 janitors put down their mops and brooms and demanded a living wage and dignity. The solidarity of many people helped the janitors win their three-week strike, which in turn sparked organizing drives in cities across the nation. As of early 2005, more than 200,000 janitors, most of them people of color, were organized into the Service Employees International Union (SEIU).

THIS ILLUSTRATION FROM *¡SÍ SE PUEDE!* HIGHLIGHTS THE POWER OF WORKING PEOPLE.

The text of the book tells the story of the strike from the perspective of the son of a single mother who is a janitor. While the book itself does not include much explicit math, it provides a context for rich lessons in the power of math.

Teachers and students can find lots of data on the bilingual (Spanish/English) Justice for Janitors website, www.seiu.org/j4j, to complement the book.

DO THE MATH

Students can work on understanding large numbers, multi-digit multiplication, problem-solving, and percentages through examining data such as:

•   Because of the strike, 8,500 janitors won a pay raise from $7.20 to $9.10 an hour. What percent increase was this? How does this compare to the rate of inflation?

- Assuming a 30-hour work week, how much money did a janitor earn (before taxes) with the old wage and with the new one? What is the difference?

- Many of the janitors are forced to work only part time and thus don't earn enough money to prevent their families from living in poverty. In 2004 the U.S. government said a family of four had to earn more than $18,850 per year to be considered above the poverty line. Even on the new salary of $9.10 an hour, if a person with a family of four was only able to work 30 hours a week, how much below the 2004 federal poverty threshold would the family be?

- In the book the mother says to her son: "Even though I work full time as a janitor, I also have to clean houses and wash clothes on the weekend. That means we don't have any time together. And I can't afford to buy the medicine Abuelita needs to help her sore bones feel better." According to the SEIU, most janitors get no healthcare benefits from their jobs. The SEIU estimates that by 2006, the cost of a family health-insurance premium is expected to reach $14,500 a year. How does that figure compare to the salary of a janitor working at $9.10 an hour for 40 hours a week? ∎

For additional information on the strike, see the Justice for Janitors website, www.seiu.org/j4j. For additional teaching ideas using this book, see www.rethinkingschools.org/math.

CHAPTER THIRTEEN

Math, Maps, and Misrepresentation

BY ERIC GUTSTEIN

What happens when students begin to question what they have been taught in school? What other questions does this help them raise, and how does it help them to better understand the world? How can teachers create these kinds of experiences?

I considered such questions as I taught my eighth-grade mathematics class at Rivera, a public school in a Mexican immigrant community in Chicago. As part of my job as a faculty member at the University of Illinois–Chicago, I taught one middle school math class at Rivera on a daily basis for a number of years. My goal as a teacher has been influenced by the work of Paulo Freire and other critical educators. I want to help my students learn to read (i.e., understand) the world — through learning and using mathematics — as a way for them to begin to write (i.e., change) the world.

Going Beyond Mathematics

As the basis of my curriculum, I used *Mathematics in Context* (MiC) an innovative mathematics curriculum developed in accordance with the National Council of Teachers of Mathematics standards. I found that MiC helped students learn to use mathematics to understand the world, as it developed many critical mathematical reasoning skills, but by itself, it was not enough. Therefore, I developed 17 real-world mathematics projects that my seventh- and eighth-grade students completed over the almost two years I was their teacher. (I moved up to eighth grade with my class.) This story, from the spring of 1999 when they were in eighth grade, is about one of those projects.

One significant lesson I learned is that going beyond mathematics is important in helping middle school students read the world with mathematics. Teachers need to develop a classroom culture that incorporates reading the world and examining injustice and oppression. An important part of going beyond mathematics is to try to normalize politically taboo topics. For example, my students and I had many conversations about race and racism, and they were central to a number of our classroom projects. I found that such an orientation is vital for students to appreciate and be more interested in mathematics, because they begin to see that mathematics can help them make sense out of their surroundings.

Analyzing Map Projections

The project I describe here was called Analyzing Map Projections — What Do They Really Show? Maps are two-dimensional representations of the earth that we often take for granted. Few of us think that our standard maps might be woefully inaccurate, and we do not often consider how the images students see everyday on classroom walls shape their perceptions of the world. Mathematics is central to map making, and different mathematical ways of representing the world produce very distinct maps.

A goal of the map project was for students to use mathematics to analyze diverse map projections and to raise questions about what the various maps showed — and why. A larger goal of this project was to help students develop a more critical outlook towards knowledge in general.

I used two very different projections: the Mercator projection (developed in 1569 in Germany), which is the traditional map in U.S. schools (including Rivera); and the Peters projection (developed in 1974 by Arno Peters). The sizes, shapes of land masses, and coloring schemes of the maps are quite distinct. The Mercator map was developed during European expansion when colonial exploitation required that maps be used to navigate accurately (so as not to repeat Columbus' blunder) and was used successfully to find new territories. All maps unfortunately are misleading because they are two-dimensional projections of a sphere — and the Mercator suffers from serious visual distortion by altering the relative size of land masses. This is because the scale changes as you move away from the equator. Thus countries far from the equator (e.g., Greenland) appear much larger than they are. (Some Mercator maps have, in fine print, an explanation of this distortion and the mathematical information necessary to find the actual areas.)

For example, Mexico is about 760,000 square miles, Alaska is about 590,000, but Alaska looks two to three times larger because it is farther from the equator. The representations of Greenland and Africa are more distorted. Greenland, at 840,000 square miles, appears roughly comparable to Africa, which, at 11,700,000 square miles, is about 14 times larger. In addition, Germany is near the center of the map, which may have made sense from the perspective of European expansionism. However, since Germany is in the northern quadrant of the earth (Berlin is 52 degrees north), the only way to make it the center is to push the equator approximately two thirds of the way down the map. This compresses the Southern Hemisphere

TERRITORY THAT MEXICO LOST TO THE UNITED STATES

MAP TEACHING SUGGESTIONS

Using this map as a guide, students can use atlases and Internet sources to measure and tabulate the total square miles of land that the United States took from Mexico following the war of 1848.

DO THE MATH

Have students calculate the percentage of Mexico that was taken and the percentage of the continental United States that was originally part of Mexico.

For the land taken, the United States paid $15 million. Students can calculate the price per acre.

VARIATION

Another option is to give the students no data other than Wyoming's area (about 100,000 square miles) and tell them to use that as a base for finding out how much land the United States took from Mexico. For deeper study, students can refer to the chapter on this war in Howard Zinn's *A People's History of the United States.*

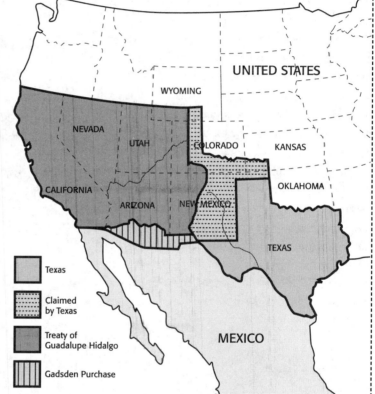

and enlarges the Northern.

These distortions remain true of today's Mercator maps, even though we now have better navigational means than the Mercator to sail across the Atlantic. Whether or not Mercator meant consciously to diminish the South, it is worthwhile to focus on the effect of the map's widespread use today.

In contrast, the Peters projection was developed to fairly and accurately portray the earth. As Ward Kaiser writes in an explanation of the Peters projection, "Peters is ... clearly focused on justice for all peoples, recognizing the values and contributions that all nations and all cultures can bring to the emerging world civilization."

The Peters map distorts shapes somewhat, but unlike the Mercator, it accurately presents the relative size of land masses.

As part of our project, I gave each group of three students a large Mercator map, borrowed from classrooms in the building. I gave them Mexico's size to use as a unit of measure and turned them loose to measure and compare several areas on their Mercator map (for example, Greenland and Africa; Mexico and Alaska). They used a variety of mathematical means to estimate the areas. Some traced Mexico and estimated how many Mexicos fit into other countries, then they multiplied the area of Mexico by the number of Mexicos that they estimated would fit into the other countries. Others overlaid centimeter grids on top of Mexico to determine the area per square centimeter, then measured other countries and found their areas. And some students used a combination of methods, including reallocating areas to make rectangular shapes whose sizes were easier to estimate. Afterwards, they dug through our almanacs to find the real areas of the places they had measured and compared those to their estimates. Finally, I had students respond in writing to several questions, both in their groups and individually.

Questioning Past Learning

My students determined for themselves that the Mercator map did not show equal areas equally. This was further confirmed when they examined the Peters map. They were astonished and upset to discover the Mercator's visual distortions, and the majority felt they had been "lied to." Many were quite concerned about the implications of this misinformation, and some questioned other sources of knowledge and things they had learned.

In particular, many wanted to know why they had been mistaught. As one of my students, Rosa, commented, "The questions raised in my mind are why teachers never told us how wrong the map was."

Another student, Lupe, took a position of advocacy: "I think it's sad that we've all been taught this way. We should make our analysis public and let it be known. I just want to understand what [is] the point exactly of Mercator's map. What did he want us to believe, to see?"

One of the questions I had asked was, "Knowing we were all raised on the Mercator map, how does that make you feel?" In response, Marisol commented that it "makes me

WORLD MAP **MERCATOR PROJECTION**

think what other wrong information we have been given since childhood. It makes you doubt your social studies book, history written by the white people." And Gloria summed up the feeling of many. "Thinking that we were all raised with the Mercator map makes me feel kind of miseducated," she said, "because all these years we've been using the wrong map."

Race as a Central Issue

As politically taboo topics became normalized in our class, race and racism became central discussion topics. Students talked and wrote about race and racism as easily as about field trips and homework. A number of my projects focused more closely than this one on race (see "Real-World Projects, page 117). However, because race was an ordinary topic of conversation and a political issue about which students were concerned, and because I asked which races (in general) lived in the North and the South, students also analyzed the project using race as a point of inquiry.

Several found the shrinking of the South relative to the North and the skewing of the relative sizes of Alaska and Mexico especially trou-blesome. Sandra had a particularly cogent analysis. "Doing this project has opened my eyes in different ways," she wrote. "I am learning how small details like maps, etc., have a lot to do with racism and power. Even though these kinds of things are small it can make a big difference in a person's view after learning what's really going on."

I also asked students to reflect on why the Mercator map is used so much in schools. One student, Elena, wrote in response, "I guess that's so because they wanted to teach [that] all Americans [are] superior and that all whites (that color and race) are better and superior than us (brown or lightly toasted, hardly white and Mexican). We were always taught that we were a minority and didn't deserve anything."

Issues of race and racism also surfaced in students' responses to the last question I asked: "In your opinion, is this in any way connected to anything else we've studied over the last two years?" Among those who answered affirmatively, Javier wrote, "It all has to do with white being the superior … because the mapmaker was from Europe, he put Europe in the 'middle of the world' and on top of other countries. But

WORLD MAP **PETERS PROJECTION**

as we know, there is no such thing as a middle of the world."

Marisol also commented on race. "I think this is connected to what we studied over the past two years about equality, minorities, etc.," she wrote. "These maps seem to make minority-filled countries smaller than white-filled countries. To make us smaller and more insignificant? Yes, definitely."

Reflections

A principal purpose of this and other projects was to get my students to think deeply about their educational experiences and what they had learned both in and out of school. I wanted my students to discover for themselves that "what you see is not always what you get." I did not want my students to take my word, nor anyone else's, without questioning. I wanted them both to use mathematics to read their worlds and to see the value of doing so — that is, to also develop a particular disposition towards using mathematics for social analysis.

My students did begin to read their world with mathematics, but clearly they were still developing in this area; for a variety of reasons, their development was uneven. They were still struggling to form their ideas about the world, and their contradictory beliefs often surfaced. Some at times tried to explain away certain things; for example, hypothesizing that we still used the Mercator because perhaps the earth had changed since 1569 and mapmaking was just catching up. This type of accommodation repeatedly appeared in class, sometimes even from students who at other times took strong positions against various forms of injustice.

It is important to note that beyond beginning to analyze society using mathematics, my students also succeeded in mathematics on conventional academic measures. All passed their standardized tests and graduated eighth grade on time, and several made it to academic magnet high schools. Although these are necessary accomplishments under the current high-stakes testing policies, they are not enough. Educator Gloria Ladson-Billings' perspective on educating African-American students also applies to working-class Latinos and other marginalized students. In her 1994 book *The Dreamkeepers,* she writes:

> [P]arents, teachers, and neighbors need to help arm African-American children with the knowledge, skills, and attitude needed to struggle successfully against oppression. These, more than test scores, more than high grade-point averages, are the critical features of education for African Americans. If students are to be equipped to struggle against racism, they need excellent skills from the basics of reading, writing, and math, to understanding history, thinking critically, solving problems, and making decisions.

A broader definition of educational success incorporates the vision of working for social justice. Mathematics can be an appropriate tool to realize this vision. The words of my student Lupe exhibit the sense of purpose needed in this struggle.

In commenting on our map project she wrote, "This is definitely connected to all we've done during the past two years. This goes back to why South America is so small in the Mercator map, to injustice and some sort of propaganda with false information. This relates to not just accepting what we have, but to search for answers to our questions. You have taught us to do that in many ways, and that only makes us grow. Who knows? Maybe we can someday prove things wrong and show the right way!" □

A New View of the World: A Handbook to the World Map: Peters Projection, W. L. Kaiser (Amherst, MA: Friendship Press, 1987). Call 800-889-5733 for ordering information.

All students' names have been changed.

REAL-WORLD PROJECTS

SEEING MATH ALL AROUND US

BY ERIC GUTSTEIN

The following is a brief, chronological list of some of the projects done by my students. Each project has a short description, a summary of the mathematics involved, and an explanation of the writing the students had to do.

1.   WHO IS THE INS ARRESTING?
Students read a newspaper article about the disproportionate number of Mexicans the INS was arresting, relative to the national-origin percentages of undocumented people. The mathematics included percentages, ratios, and proportions, and students also had to pose their own mathematical problems based on the article. Students had to write about their view of the INS policy.

2.   "MORNINGSIDE PARK TO BE PAVED FOR CONDO PARKING LOT."
Students read a newspaper article about a developer who wanted city permission to raze a small park in order to build parking for his condominium development. They had to use two maps with different scales to determine distance and time between various locations. The mathematics also included ratios and measurement. Students had to write about whether the condo developer should be allowed to raze the park.

3.   GROWTH OF LATINO POPULATION IN THE METROPOLITAN AREA.
Students read a newspaper article and had to use percentages and graphs to answer (and pose) several questions. They had to write about the social implications of the changing demographics.

4.   "TOMATO PICKERS TAKE ON THE GROWERS."
Students read a New York Times article by this name, about how farmworkers' wages had gone down over 20 years, even without taking inflation into account. Mathematical ideas included COLA, absolute/relative comparisons, percentages, graphing, and inflation-adjustments. Students had to write about what they would do if they were in the shoes of the workers.

5.   DEVELOPING SURVEYS (YEAR 1).
Students developed and conducted surveys in small groups on mean-

117

ingful topics (i.e., serious and non-trivial). These included abortion (two groups), gay marriage, Darwin's theory of evolution, teen pregnancy, and discrimination against Latinos. Each group had to survey 80 to 100 people, make graphs, present findings, propose open research questions, write conclusions, and justify their choices along the way.

6. WILL DEVELOPMENT BURY THE BARRIO?

Students read a long feature article about development in the school neighborhood. The mathematics included percentages, compound interest, ratios, and proportions. Students had to write about whether or not development would destroy the community.

7. RACISM IN HOUSING DATA? (PART 1).

Students were given the 1997 median house price of the most expensive suburb in the area and asked if racism was a factor. Students had to describe and hypothesize data that would lead them to conclude that racism was a factor and data that would lead them to conclude racism was not a factor. Mathematics included data analysis, percentages, and problem-posing.

8. SAT DATA.

Students were given 1997 SAT and ACT scores, disaggregated by race, gender, and income level. Students had to analyze correlations, create scatterplots, and write a letter to the Educational Testing Service raising questions and/or giving their opinions.

9. WORLD WEALTH SIMULATION.

Students used data on wealth and population by continent to distribute cookies ("wealth") to people in the classroom ("continents"). Mathematics included determining percentages, absolute/relative comparisons, ratios, and proportions. Students had to write their thoughts on the distribution. (For a complete description of this activity, see page 64.)

10. "WE ARE NOT A MINORITY."

Students looked at a photo of a billboard which showed Che Guevara with this statement. They had to write about how mathematics entered into the picture. After writing, the class analyzed world population by race and discovered that people of color constitute a large majority of the world's population.

11. WATCH AD.

Students read a newspaper advertisement for a watch that cost $26,125. They had to write a creative piece about the watch and its price, using mathematics at the level of at least eighth grade, and have their opinion about the watch and ad come through in the piece.

12. FAIR ADVERTISING (BARNES & NOBLE AD).

Students read a newspaper advertisement which used graphical representations to show the volume of books owned by Barnes & Nobles and a competitor. Students had to define a "fair" ad, analyze the advertisement to determine whether it was fair, justify their decisions, and then, if they believed the ad to be unfair, create a new ad that was fair. Mathematics included measurement, geometry, percentages, ratios, proportions, and graphing.

13. U.S. WEALTH SIMULATION.

Students used data on the wealthiest one percent, next 19 percent, and bottom 80 percent of the U.S. population to distribute cookies to themselves (representing the three groupings) by the wealth percentage for each group. Mathematics included percentages, absolute and relative comparisons, ratios, and data analysis. Students had to write their thoughts on the U.S. wealth distribution and how it compared to the world wealth distribution, and also whether using mathematics was necessary for them to understand the ideas. (For a description of a similar activity, see page 68.)

14. DEVELOPING SURVEYS (YEAR 2).

Similar to year 1 but in greater depth.

15. PETERS VS. MERCATOR MAP PROJECTIONS.

Students were given Mercator maps and the area for Mexico and had to find the areas of several other land masses. They then consulted almanacs to determine the real areas, which demonstrated that the visual representation of the Mercator projection were inaccurate. They then compared and contrasted the Peters and Mercator maps. Mathematics included measurement, geometry, ratios, and proportions. (For a complete description of this activity, see page 111.)

16. RACISM IN HOUSING DATA? (PART 2).

Students analyzed data from almost 200 metropolitan suburbs to correlate house prices with racial composition of the suburb. Mathematics included data analysis and graphing. They had to graph, use scatterplots, find correlations, and then write and justify whether or not they believed racism was a factor in their data and, if so, how and why.

17. THE MATHEMATICS OF RANDOM DRUG TESTING.
Students simulated a situation where school officials had a very accurate drug test and a small percentage of students were using drugs. They then looked at the probability that a randomly chosen student who tested positive was really using drugs. Adapted from the IMP curriculum. Mathematics included probability, expected value, randomness, and statistical reliability.

18. DRIVING WHILE BROWN/DRIVING WHILE BLACK.
Students simulated racial profiling data of "random" traffic stops, then compared the results of their simulations to actual data. Students analyzed the data to decide whether racial profiling was a real problem and, if they believed it was, proposed what should be done about it. Mathematics included probability, data analysis, and rational numbers. (For a complete description of this activity, see page 16.)

19. B2 BOMBER PROJECT.
Students compared the cost of one B2 bomber to the cost of four-year scholarships to a prestigious out-of-state university for all the graduating seniors in the neighborhood high school, as well as all of Chicago. Mathematics included using large numbers, scientific notation, and ratios. (See the related activity, page 76.)

20. MATHEMATICS OF DEMOCRACY.
Students read a cartoon (at right) about public opinion on the Iraq War. They then wrote articles or editorials about the cartoon and analyzed how the cartoonist used mathematics to make his point. Mathematics included percentages, graphs, ratios, and proportions.

21. MORTGAGE LOANS — IS RACISM A FACTOR?
Students read a newspaper article documenting the disparity between African Americans, Latinos, and whites in securing home loans in the Chicago area. Students analyzed the data, solved a number of problems, and wrote essays arguing whether racism was a factor in the disparities and using mathematics to justify their views. Mathematics included algebra, percentages, graphs, data analysis, and ratios. (For a complete description of this activity, see page 47.) ■

JIM BORGMAN (c) Reprinted with special permission of King Features Syndicate.

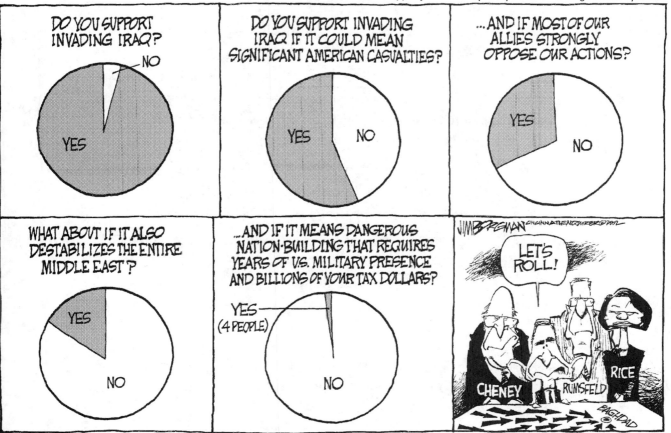

DO YOU SUPPORT INVADING IRAQ?

CARTOON TEACHING SUGGESTIONS

Show students the cartoon and ask, "In the cartoon, why did the responses to the survey change? What does this tell us about the importance of how questions are constructed?

What does the cartoon imply about President Bush's concern for public opinion? See Mathematics of Democray in "Real-World Projects," at left, for other ways to use this cartoon.

DECONSTRUCTING BARBIE

MATH AND POPULAR CULTURE

BY SWAPNA MUKHOPADHYAY

Math and popular culture are rarely associated, although phrases like "do the math" are commonly heard. Undoubtedly this is a prime example of the unfortunate disconnect between math in school and its real-life applications. Many teachers, however, use mathematics for the analysis of complex real-world issues. Teaching mathematics as a cognitive tool for everyday sense-making is an approach that addresses the alienation of many students from early grades.

Beyond sense-making, using mathematics as a tool to interrogate issues of importance to students, their communities, and society in general brings to light the generally hidden cultural, historical, and political nature of mathematics education.

Here's an activity that treats mathematics as a tool for learning about a number of social issues. It begins with inviting a Barbie doll — the popular cultural icon — to the classroom. The activity teaches concepts such as averages and ratios, in a way suitable for middle school students.

MATERIALS

Besides a few Barbie dolls from my thrift-store collection, I also bring tape measures, a ball of string, and calculators as my material props.

PROCEDURE

The task starts with a probe: What would Barbie look like if she were as big as you? Students, working in small groups, need to figure out first the "average" of the group, so that they can begin to "construct" a real-life Barbie of the same height. After confronting the mathematical notion of average and its contradiction to real life (for example, "Jenny," the "average" in our group, is a blond European American, whereas I am a dark-skinned Indian), the students measure the ratio of the height of the designated person to that of the Barbie doll to find the enlargement ratio or the scaling factor.

Then, with a table I provide, they measure the body parts of the doll and compute its corresponding real-life measurements.

The next part of the task consists of drawing the contour of Jenny-the-average on butcher paper with one color marker, and then superimposing the blown-up Barbie on Jenny's image with a different color. Drawing two human figures — the contour of familiar Jenny and Barbie-as big-as-Jenny — immediately highlights the unrealistic

DECONSTRUCTING BARBIE

MEASURE UP, BARBIE!

$$\text{scaling factor} = \frac{\text{height of person}}{\text{height of Barbie}}$$

BODY MEASUREMENTS

	BARBIE	BARBIE SCALED	REAL PERSON	RATIO
BUST				
WAIST				
HIPS				

OUTLINE MEASUREMENTS

	BARBIE	BARBIE SCALED	REAL PERSON	RATIO
LEGS				
ARMS				
HEAD				
NECK				

dimensions of Barbie, which are not so obvious on the doll itself, or even in the computed measures collected in tabular form.

Repeating the exercise with male superhero action figures and the "average" male student makes it clear that the issue is not confined to female representations.

DISCUSSION

This activity then generates discussion connecting to the impact of representations in popular culture on body image, self-worth, and eating disorders. Other issues raised include the superficially "multicultural" nature of contemporary Barbie and the sweatshop labor that produces the dolls. ∎

CHAPTER FOURTEEN

Multicultural Math

One Road to the Goal of Mathematics for All

BY CLAUDIA ZASLAVSKY

As I look back on my teaching experiences in Greenburgh Central Seven, a school district just north of New York City that was known nationwide for its decision in 1951 to integrate its schools through busing, I am struck by the similarity between the problems we encountered in the 1960s and 1970s and those of today.

The solutions we worked out may be of value to teachers trying to reach all students with mathematics. In particular, this article will deal with the introduction of elements of a multicultural, anti-racist mathematics curriculum in the middle and secondary grades.

The National Council of Teachers of Mathematics recognizes the relevance of culture, saying: "Mathematics is one of the greatest cultural and intellectual achievements of humankind, and citizens should develop an appreciation and understanding of that achievement, including its aesthetic and even recreational aspects." I agree. Multicultural education should include the contributions of all peoples, as well as concern for their problems and difficulties. Further, it should embody the consideration of the factors in our society that prevent the effective participation of all members of society.

The '60s was a period of ferment and change. One effect on our school district was the students' demand for courses in African history and Swahili, courses that were relevant to their culture. The faculty, too, was offered an optional course in African history, a subject that generally had received inadequate, if any, treatment in the college curriculum.

For my term paper in the course, I decided to research the development of mathematics in Africa south of the Sahara, but I was amazed to discover that there was no heading in the library catalogs for African mathematics or for any of its branches.

Mathematical practices and concepts arose out of the needs and interests of people in all societies, in all parts of the world, in all eras. All peoples have invented mathematical ideas to deal with such activities as counting, measuring, locating, designing, and, yes, playing, with corresponding vocabulary and symbols to communicate their ideas to others. I was convinced that African peoples were no exception and that such information must be available. My research eventually led to the publication of a book on the subject: *Africa Counts: Number and Pattern in African Cultures* (see Resources, page 168).

A New Secondary School Mathematics Curriculum

Although our secondary school had few dropouts — partly because of the availability of an adequate number of guidance counselors, who stayed with the same students throughout their high school careers — we found that many students were doing poorly in the traditional academic mathematics courses and were dropping the subject as soon as they had fulfilled the minimum requirements. These students were predominantly of working-class backgrounds, both black and white. This was not an acceptable development in a district that prided itself on offering the best possible education to all its students.

Like today, the '60s was a period of great concern about the status of mathematics education in this country. The Soviet "conquest" of space with the launching of Sputnik in 1957 sent shock waves through our government and business establishment, resulting in increased funding to upgrade mathematics and science education. Our district, ever on the alert for such opportunities, obtained funds to write a new curriculum to attract potential math dropouts. We were fortunate, too, in that the community supported us as we increased the number of courses offered by the mathematics department, a number exceeding what was made available in some larger and wealthier districts nearby.

Our mathematics department developed a complete curriculum for grades nine to 12, designed to be user-friendly and combining hands-on activities with many of the topics usually covered in grades nine to 11. The 12th-grade unit was based on elementary probability and statistics. The entire curriculum included more computation than is usually found in academic mathematics courses at this level, presented in a manner that would enable students to understand the underlying concepts and some aspects of number theory. As much as possible, applications were derived from local issues and the real world.

I will describe briefly the most relevant topics in the 12th-year course, then discuss more fully one of the most successful projects, involving census data for the local area. The topics listed below were not necessarily included every time the course was given: Flexibility was a great advantage in designing our own courses. We could add what seemed relevant at the time and omit the topics that did not work well. The topics included the following:

- Probability theory, involving experiments with coins, dice, thumbtacks, and playing cards and a discussion of mortality tables, lotteries, and heredity.
- Statistical distributions based on students' data: heights, birthdays, number of children in the family.
- Descriptive statistics (tables and graphs) relating to such topics as the integration of whites and "nonwhites" in housing; the amount of nicotine in different brands of cigarettes; the lifetime dollar worth of an education; the comparison of education and earnings of males and females; and the comparison of blacks and whites with regard to education, income, infant mortality, life expectancy, and other factors.
- Computation, significance, and applications of the mean, median, and mode; histograms and the normal distribution curve.
- Computation of grade-point averages, the consumer price index, and the cost-of-living increase incorporated into the contract of the United Automobile Workers union.

Certain topics evoked a great deal of interest. This was the period of the Vietnam War, and young men were very concerned about their futures. A Department of Defense publication showed that in 1967 black men composed nine percent of the armed forces, but 11 percent of all men assigned to Vietnam and 15 percent of the deaths in Vietnam. The students discussed the reasons for the discrepancy, illustrating their arguments with stories about friends and relatives.

Another interesting set of statistics involved the difference between the cost of a shopping basket of food in the low-income and middle-class neighborhoods of New York City: Poor people paid more for their food than higher-income people! Langston Hughes' lines about Harlem from his poem "The Panther and the Leash" were appropriate: "Uptown on Lenox Avenue / Where a nickel costs a dime...."

Analysis of Data About the Community

The most empowering and, therefore, the most successful lessons were related to the students' own community. Using the Department of Commerce publication *Standard Metropolitan Statistical Areas Outside of New York City, 1960 Census,* I drew a map showing the five standard metropolitan statistical areas (SMSAs) that composed the school district. I leafed through the book, selecting the features that I thought would be of greatest interest to the students, and copied the data for each SMSA. Among the categories I used were these:

- Total population, classified as white, black, other.
- Education and median family income for each category.
- Male and female employment and unemployment rates for whites and nonwhites.
- Housing — owned or rented, condition, value; number of persons per room — for whites and nonwhites.
- Automobiles owned.

Every student received a copy of the map and the data. After some discussion, the students paired off, each couple taking responsibility for comparing the five SMSAs from the point of view of two or three characteristics. We had some general discussion about making tables and graphs to display the data, and then they got to work. Although they worked in pairs, each student was responsible for producing the tables and graphs.

(I had assumed that this was a simple assignment, but I was wrong. For some, planning the scale for the graph was a formidable task. Their first attempts revealed how little they understood. Their scales did not fit on the paper. They did not know how to label the intervals after they had marked them on the vertical axis;

if the data gave median incomes of $2,320, $4,350, $5,489, $4,190, and $3,085 for the five SMSAs, those were the numbers placed next to the marks on the vertical line. Several students were successful only after three or four attempts.)

Then the class came together to compare and analyze their productions. With the boundaries of each SMSA clearly marked on the map, each student knew exactly where he or she fit into the total picture. It was evident that the predominantly white, middle-class population had far higher incomes, better housing, more education, and all the other attributes of a higher standard of living than the more poorly endowed black and working-class population. Of course, the students had been aware of these differences before. But here were the numbers, and numbers don't lie. Here were the graphs, and a picture is worth a thousand words.

Most of the students were taking sociology as their senior-year social studies course. The social studies department considered me an honorary member of their department, and they welcomed the input from my classes as a basis for further discussion in the sociology course. Nor did parents object to our tackling such a controversial subject as societal inequities. In a district that had opted for school desegregation, we could generally count on the support of the majority of the community in dealing with such topics. Many families, both black and white, had chosen to live in this community precisely because it was less segregated than most of the surrounding area.

Both the faculty and the students considered these courses successful in overcoming math anxiety and math avoidance. We tried to eliminate the need for remedial courses at the high school and college levels by incorporating a review and by reteaching, if necessary, concepts from arithmetic in a context that enhanced

the students' self-esteem, rather than labeling them "dumb" and increasing their fear of mathematics. Furthermore, these topics in arithmetic took on more meaning than if they had been studied and drilled out of context.

Many of these potential dropouts from mathematics continued to take mathematics courses throughout their high school years. We

> ## Mathematics comes alive when students participate in activities that illustrate how mathematical decisions arose from the basic needs of societies.

were particularly successful with blacks and working class young people of all races, those groups that had been underserved and underrepresented in mathematics. Young women were not a problem; females and males in our district were equally represented in advanced mathematics classes as far as the pre-calculus level. But we did make conscious efforts to encourage all students, especially "minorities," to pursue mathematics.

The Mathematics of Various Cultures

In the early '70s I began to explore the possibilities of introducing into the curriculum the study of mathematics as a cultural product, the kinds of mathematics developed by various societies to satisfy their own needs. My focus at the time was on societies in Africa south of the Sahara; later I broadened my scope to include some of the underrepresented societies throughout the world. In recent years the mathematical and scientific developments of non-European cultures have become the focus of study for a number of scholars. Not to include such contributions is to imply that these people had no mathematics or science. The same applies to women of all races and ethnic backgrounds.

I took advantage of a sabbatical year to develop and test materials for several grade levels, based on the infusion of African mathematical ideas into the curriculum.

In our district the ninth-grade social studies curriculum included the study of Africa. The two teachers who team-taught this large, hetero-

From *Africa Counts: Number and Pattern in African Cultures*

A DIAGRAM OF A CHOKWE NETWORK SIMILAR TO THOSE DRAWN IN SAND.

geneous social studies class were perfectly happy to allow me to take over for a week. On the first day I showed slides illustrating mathematical practices that several African peoples had developed in the course of their everyday activities. The following day I introduced a unit about networks, on which we spent the remainder of the week.

A network consists of a set of points (vertices) and the paths (edges) connecting them. My favorite network is a figure that was drawn in the sand by elders of the Chokwe people of Angola and Zaire to illustrate their myth about the beginning of the world (see diagram above).

These African networks are examples of mathematical graphs. Graph theory is a growing field of mathematics and has important applications in modern technology relating to communications, highways, and so on. The task I set the students was to trace the given networks and

to determine which designs were traceable: that is, could be drawn in one sweep of the pencil, without either lifting the pencil or going over the same line segment more than once. Then I challenged the students to determine the necessary conditions for traceability.

The project was open-ended; some students were able to trace only the given networks; others drew traceable networks of their own invention, and still others tried to respond to the theoretical challenge. Most students indicated on their evaluation sheets that this was a positive experience for all, including the teachers, who later told me that several of these students had not otherwise participated in class since the beginning of the school year. They had rejected the papers I distributed the first day, but when they saw their classmates engaged in hands-on work and animated discussions, they, too, wanted to participate.

It seemed to me that, in general, this infusion of the cultural aspects into the curriculum was successful. African-American students, in particular, found such materials relevant to their lives and backgrounds. One young woman chose to instruct an eighth-grade class in African applications of network theory as her senior year project.

Many mathematical topics lend themselves to the infusion of cultural applications. In the process of designing a rug similar to those woven by Navajo women, students deal with symmetry, geometry, and measurement. At the same time they learn to appreciate the degree of mathematical knowledge required to weave complex designs, generally with no pattern and no specific instructions to follow. With this understanding, "women's work" takes on new significance. To mention just one problem: How does the weaver compute the number of rows required to make the border pattern "come out right" when she reaches the far side of the rug?

Mathematics comes alive when students participate in activities that illustrate how mathematical decisions arose from the basic needs of societies. For example, why do people build their homes in certain shapes and sizes and use particular materials? An investigation into the building styles in various cultures is a valuable source of experiences with shapes and sizes, perimeter and area, estimation and approximation, while at the same time it shows the relevance of mathematics to social studies, art, and other subjects. Students might consider a tipi, an African mud-and-wattle roundhouse, or an Inuit igloo a "primitive" dwelling compared with an urban apartment house or suburban ranch house. Yet the people who build these homes, or built them in years past, are using their available materials and technology to the best advantage.

Relevant, Not Trivial, Multiculturalism

A word of caution: Teachers must be careful that they do not introduce cultural applications as examples of "quaint customs" or "primitive practices." These applications must form an integral part of the mathematics curriculum. They must inspire students to think critically about the reasons for these practices and to dig deeply into the lives and environment of the people involved. It is so easy to trivialize the concept of multicultural education by throwing in a few examples as holidays approach. Better not to do it at all!

The multicultural context is relevant to many aspects of the mathematics curriculum. A discussion of the number words and numeration systems of non-English-speaking peoples may do wonders in raising the self-esteem of students who speak these languages, while also enhancing the understanding of all students. It may come as a surprise that in some languages, numeric grouping is done by 20s rather than by 10s as in English. Games of chance, games of skill, and patterns in art and architecture are all sources of learning experiences. Some of the richest contributions may come from students in the class and their families.

Bringing the world into the mathematics class by introducing both cultural applications and current societal issues motivates and empowers students. Such mathematical content offers opportunities for project work, cooperative learning, connections with other subject areas, and community involvement. To carry out such a program effectively requires a new approach to curriculum development, teacher education, and assessment processes. But it is well worth the effort. □

This article is adapted from a version that originally appeared in *Reaching All Students with Mathematics,* Gilbert Cuevas and Mark Driscoll, eds. (NCTM, 1993). For a list of all references and citations that accompanied the original article, see www.rethinkingschools.org/math.

TRACKING PA ANNOUNCEMENTS

COLLECTING AND ANALYZING DATA IN THE CLASSROOM

BY ROBERT M. BERKMAN

Several years ago I was hired to teach in a public middle school located in a blue-collar section of Brooklyn. There was an almost constant flow of announcements over the public address system, to the point where a 40-minute class might be interrupted half a dozen times.

Fueled by defiance and exasperation, I decided to reserve a corner of my chalk-board to tick off each interruption. My students were confused when I abruptly and silently stopped a lesson to add the latest announcement to my tally, but after the third or fourth time, they quickly put together the cause and effect. Their reaction was to chuckle. Apparently they were so conditioned to this constant stream of blather that they were surprised that it bothered me at all. I remember one student laughing as I counted off what must have been the 10th announcement of that day, remarking: "Yo, Mr. Berkman, that sh–– be driving you crazy!"

COLLECTING THE DATA

After collecting a week's worth of tally marks, my students were not surprised that the total was quite high (as many as 25 announcements were made in a single day). Yet they felt that what I was doing had grown, to use their word,

"lame." I told them that I had a theory that there were certain times when the PA system was most used, and that perhaps I could prepare myself for those times (by not attempting to actually teach a lesson).

A spirited discussion ensued: One student said it was always at the end of the day, when kids had to leave early, while others thought it was in the morning, when the principal delivered his daily pep talk. I refined my corner of the board to include eight boxes, each representing a different period of the day.

That week my students were careful to make sure I kept track of all announcements, popping in between periods to ask: "Hey, Mr. Berkman, I heard two last period. Did you get them?" We looked at the daily totals, eventually creating a histogram for the entire week. Our results showed that most announcements occurred during the middle of the day, in fact, most often during lunch periods, when teachers were either being reminded to show up at lunch duty or kids were being sent to the office for some trouble they had caused.

During the next few weeks, we kept track of announcements in a variety of ways. We set up a chart classifying announcements by topic, including "general interest," "requests for a teacher," "requests for a student," and "miscellaneous." Our research found that while "general interest" announcements occurred at the beginning and end of the day, the "student requests" happened during lunch periods, while "teacher requests" and "miscellaneous" occured at random times.

HISTOGRAMS AND PIE CHARTS

By this time, the students were collecting their own data, to the point that they would correct the information I had carefully recorded. After a discussion about the announcements, the students grumbled about the different people who made them, noting that it was usually the principal, his secretary, or one of the three assistant principals who was the "guilty" party. To test their theory, my students kept charts on who was making announcements, what time of day they were made, and the subject of each. They tabulated this data and made a variety of histograms, pie charts, and line graphs.

When the students asked what we would do with all the information, I jokingly suggested that we write up our conclusions and submit the results of our investigation to the principal in an effort to make him aware of how many interruptions were taking place during the day. As a new teacher, I didn't seriously consider taking this action. It was hardly necessary: Within a few days, other teachers had noticed their students tracking announcements in their notebooks, which soon came to the attention of the assistant principals.

TAKING ON SCHOOL CULTURE

During the next week the steady flow of interruptions slowed to a trickle — although they did stage a comeback after a month or two. (Was this a sociological example of the Heisenberg Uncertainty Principle, that the act of recording these interruptions actually influenced how many were made?)

I cite this experience because what began as an angry rejection of school culture was transformed into a subject for data collection, evaluation, and analysis. The students took interest in this because it was something that annoyed them day in and day out, but they didn't regard it as something that was worthy of collection and study, for they were so enmeshed in the school culture that it didn't occur to them to question it.

Collecting data got them involved in thinking about why things were the way they were, and as they answered one question, others sprang up, which ultimately led them to the realization that things didn't have to be this way: Our school could function just as effectively without all these interruptions; furthermore, it was the adults who had to mend their ways, not the kids. ∎

Infusing Social Justice Math into Other Curricular Areas

CHAPTER FIFTEEN

Ten Chairs of Inequality

BY POLLY KELLOGG

---

nequalities of wealth are becoming more extreme in the United States. While billionaires double their wealth every three to five years, we have by far the highest poverty rate in the industrialized world. No industrialized country has a more skewed distribution of wealth. Students need information about this concentration of wealth — and the power that accompanies it — in order to become critical thinkers and aware citizens.

A Boston-based group, United for a Fair Economy, has developed a simulation activity to dramatize the increasingly unequal distribution of wealth. I describe here how my college human relations classes respond to the exercise. It can easily be adapted for younger students.

To begin the simulation, I ask 10 students to volunteer to line up at the front of the room, seated in their chairs and facing the rest of the class. I explain that each chair represents 10 percent of the wealth in the United States and each occupant represents 10 percent of the population; thus when each chair is occupied by one student, the wealth is evenly distributed. I explain that wealth is what you own: your stereo, the part of your house and car that are paid off, savings like stocks and bonds, vacation homes, any companies you own, your yachts, villas on the Riviera, private jet airplanes, etc. Then I ask students to estimate how much wealth each family would have if the wealth were equally distributed. Students usually guess about $50,000 and are surprised to hear that the answer is $250,000.

I ask them what it would feel like if every family could have a $100,000 home, a $10,000 car paid for and $140,000 in savings. Some make

their workers." Sometimes a student, often female, will say, "I don't like to be above other people."

I invite the class to speculate how many chairs belong to the richest student, whom I will call Sue. In 1998, the richest 10 percent owned 71 percent of the wealth — thus Sue controls about seven chairs. I tell the six students sitting nearest to Sue to give up their chairs to her and move to the poorer end of the lineup. I provoke them by telling them that the standing students can sit on the laps of the three students seated at the end, and I invite Sue to sit in the middle of her seven chairs, to stretch out, relax, or even lie down across the chairs.

comments like, "It'd be wonderful. I wouldn't have to work two jobs and take out a loan to go to college." But many can't imagine such a society. Others express concern that the incentive to work would be taken away: "It sounds like socialism."

I tell those worried about socialism that they have nothing to fear. We have nowhere near an equal distribution of wealth in this country. The poorest 20 percent of the population is in debt, and the next 30 percent averages only $5,000 in wealth (primarily in home equity).

I ask students at either end of the lineup which one of them wants to represent the richest 10 percent and experience being rich. Some students volunteer happily and others express distaste at the idea. When asked about their motives, they say, "I'll never be rich, so I'd like to see what it feels like," or "I don't want to oppress other people, and rich people exploit

I then announce that Sue's arm represents the wealthiest one percent of families and that her arm's share of the wealth doubled from two chairs (22 percent of the wealth) to four chairs (38 percent) from 1979 to 1998. I solicitously help Sue find a comfortable position with one arm stretched over four chairs. To engage Sue in clowning and playing up her role, I offer her food or drink.

"I Want a Revolution"

I ask the other nine students crowded around three chairs what life is like at their end of the line. "We're pissed and tired of working all the time," is a typical comment. Another is, "I want a revolution." I ask students if, in real life, they or people they know are crowded into the bottom one or two chairs, and what that's like. Working-class students tell stories of financial stress they have experienced, such as, "My mother had to work at two jobs to support us." "My family was really poor when my dad was laid off. We lived on macaroni and cheese." Often one student, usually a white male, says he has hopes that he can work hard and join Sue.

Students' knowledge of how inequality is rationalized erupts when I ask, "What do those in power tell us to justify this dramatic inequality?" Typical student answers are: "They work harder than we do." "They create jobs." "The United States stands for equality and justice for all." "It's our fault if we don't make it." If students do not mention scapegoating, I bring it up. I may select one student to represent the poorest 10 percent and ask, "Wouldn't there be more money for the rest of you if he or she weren't ripping off the system for welfare?" I also ask, "Who does Sue want you to blame for your tough economic conditions?" Answers range from welfare moms and immigrants, to gays and lesbians, to bad schools.

When I ask the nine students grouped around the three chairs why they don't get organized to force a redistribution of the wealth, they offer a variety of answers. "We're too busy working to organize." "We are told we can't

change things." "We don't get along with each other." "They'd call out the army to stop us."

At some point I ask the class to describe the "super rich" — the one percent, Sue's arm. College students share examples from their experiences. A junior high coach described a local billionaire who offered to write the coach's school a check of any amount in order to get his child on the baseball team. Another student worked as a waiter in an elite club where "you had to be elected to be able to have lunch there." The club was all white and only recently began allowing women on the premises; some of the older men refused to let her serve them because they resented her presence. Another student described doing carpentry in a mansion of the DuPonts, "The faucet in the kid's bathroom cost $3,000."

The athletes and entertainers whose salaries are hyped in the media and newly rich entrepreneurs like Donald Trump and Bill Gates are always mentioned. I point out that these are the upwardly mobile people, who moved from the three chairs up to Sue's chair. How often does this happen? Why do we hear so much about them? I want students to understand that the exaggerated publicity about these rags-to-riches icons perpetuates the myth that anyone who tries can make it.

Most texts and teachers stop after they have taught students about the unequal distribution of wealth, but that is only a piece of the picture. We need to go on to ask why wealth is so unequally dis-

Illustration: United for a Fair Economy

tributed. Where does wealth come from? Why does our system concentrate wealth in the hands of so few? And what can ordinary people do to effect change? The simulation creates a foundation for these later lessons. □

FAST FOODS, JUNK FOODS, AND ADVERTISING

ANALYZING THE BARRAGE OF ADVERTISING AIMED AT CHILDREN

BY BOB PETERSON

In 2001 the U.S. Surgeon General announced that obesity had reached epidemic levels in this country. This epidemic includes children. While there are several causes of this epidemic, one clear aspect is the constant barrage of advertising students receive encouraging them to eat junk food.

Juliet Schor, in her book *Born to Buy*, provides a wealth of data and analysis of what she calls the "corporate-constructed childhood." Her book is filled with data that could be used by math teachers to develop lessons that students would directly relate to and benefit from. Two of those lessons are included here.

LESSON 1

A 1999 study from the University of Delaware found that 63 percent of the 353 ads shown on Saturday morning children's television were for food products, mostly high-sugar foods. The authors wrote, "There have been no food advertisements for fruits and vegetables … in the past 25 years."

ACTIVITY

As a homework assignment, teachers can provide students with a chart for tabulating, in a set period of time, the number of commercials, product types, and brand names they are exposed to. Together the class can combine the data or do a separate analysis. Depending on the level, students could make charts and graphs, convert data to percentages and fractions, and contrast their findings with national data.

LESSON 2

According to Schor, "An accounting from the May 2002 issue of Nickelodeon magazine, the most popular kid magazine in the country, found that of its 24 ad pages, 18 and a half were for junk foods. Five and a half were for candy and gum, and the remaining 13 were devoted to products such as Pop Tarts, Tang, Jell-O, and sugary cereals."

ACTIVITY

Teachers can locate popular children's magazines and have students tabulate, graph, and analyze the nature of advertisements in those magazines. Again depending on the math level, students can display their data in different ways.

QUESTIONS FOR CLASS DISCUSSION

1.   What are children learning from these advertisements?

2.   Why aren't there advertisements/commercials for fruits and vegetables?

3.   What influence do you think these advertisements have on your younger brothers and sisters? What would happen if your brothers and sisters only followed the advice of these advertisements?

4.   In Sweden, these kinds of advertisements are against the law, as people there think that children are too easily influenced to buy things that are unhealthy for them. Do you think that such advertisements should continue here in this country? Why or why not?

5.   How does using math help us better understand what is happening in our community, on TV, in magazines, and to our health?

OTHER ACTIVITIES

•   Students could research the nutritional content of some of the advertised products.

•   Students could develop presentations on the impact of such advertising and present them to younger students, other teachers, or parent organizations.

•   Students could imagine what would happen if a person just followed the advice of the advertisers, and draw and write about this idea. Where appropriate, teachers could show selections of the documentary movie *Supersize Me* by Morgan Spurlock, which follows Spurlock's misadventures as he endeavors to eat only McDonald's food. ■

CHAPTER SIXTEEN

Write the Truth: Presidents and Slaves

BY BOB PETERSON

During a lesson about George Washington and the American Revolution, I explained to my fifth graders that Washington owned 317 slaves. One student added that Thomas Jefferson also was a slave owner. And then, in part to be funny and in part expressing anger — over vote fraud involving African Americans in the then-recent 2000 election and the U.S. Supreme Court's subsequent delivery of the presidency to George W. Bush — one of my students shouted: "Bush is a slave owner, too!"

"No, Bush doesn't own slaves," I calmly explained. "Slavery was finally ended in this country in 1865."

Short exchanges such as this often pass quickly and we move onto another topic. But this time one student asked: "Well, which presidents were slave owners?"

She had me stumped. "That's a good question," I said. "I don't know."

Thus began a combined social studies, math, and language arts project in which I learned along with my students and which culminated in a fascinating exchange between my students and the publishers of their U.S. history textbook.

After I admitted that I had no clue exactly how many presidents owned slaves, I threw the challenge back to the students. "How can we find out?" I asked.

"Look in a history book," said one. "Check the Internet," added another.

I realized that I had entered one of those "teachable moments" when students show genuine interest in exploring a particular topic. Yet I had few materials about presidents and slaves and no immediate idea of how to engage 25 students on the subject.

I also recognized that this was a great opportunity to create my own curriculum, which might help students look critically at texts while encouraging their active participation in doing meaningful research. Such an approach stands in sharp contrast to the "memorize the presidents" instruction that I suffered through growing up and that too many students probably still endure. I seized the opportunity.

Action Research by Students

First, I had a student write the question "Which presidents were slave owners?" in our class notebook, "Questions We Have." I then suggested that a few students form an "action research group," which in my classroom means an ad hoc group of interested students researching a topic and then doing something with what they learn. I asked for volunteers willing to work during recess. Several boys raised their hands, surprising me because I would have guessed that some of them would have much preferred going outside to staying indoors conducting research.

At recess time, Raul and Edwin were immediately in my face. "When are we going to start the action research on the slave presidents?" they demanded. I told them to look in the back of our school dictionaries for a list of U.S. presidents while I got out some large construction paper.

The student researchers were excited to present their findings and decided to do so as part of a math class.

The dictionaries, like our social studies text, had little pictures of each president with some basic information. "Why doesn't it show Clinton?" Edwin commented. "He's been president forever."

I think, yeah, Clinton's been president four fifths of this 10-year-old's life. But I kept that thought to myself and instead replied, "The book is old."

"Why don't they just tell whether they have slaves here in this list of presidents?" asked Edwin. "They tell other things about presidents."

"Good question," I said. "Why do you think they don't tell?"

"I don't know, probably because they don't know themselves."

"Maybe so," I responded. "Here's what I'd like you to do. Since slavery was abolished when Lincoln was president, and since he was the 16th president, draw 16 lines equal distance from each other and list all the presidents from Washington to Lincoln, and then make a yes-or-no column, so we can check off whether they owned slaves." (See page 148 for a sample chart.)

I was soon to find out that filling in those columns was easier said than done.

When my students and I began investigating which presidents owned slaves, our attempts focused on traditional history textbooks and student-friendly websites from the White House and the Smithsonian Institution. These efforts turned up virtually nothing.

We then pursued two different sources of information: history books written for adults and more in-depth websites.

I brought in two books that were somewhat helpful: James Loewen's *Lies My Teacher Told Me* and Kenneth O'Reilly's *Nixon's Piano: Presidents and Racial Politics from Washington to Clinton*. By using the indexes and reading the text out loud, we uncovered facts about some of the presidents.

We also used the web search engines Google and AltaVista and searched on the words "presidents" and "slavery." We soon learned we had to be more specific and include the president's name and "slavery" — for example, "President George Washington" and "slavery." Some results were student-friendly, such as the mention of Washington's slaves (and some of their escapes) at www.mountvernon.org/education/slavery. There was also a bill of sale for a slave signed by Dolley Madison, the wife of president James Madison (for a link to the document, see www.rethinkingschools.org/math).

Many websites had a large amount of text and were beyond the reading level of many of my students. So I cut and pasted long articles into word-processing documents so we could search for the word "slave" to see if there was any specific mention of slave ownership.

In their research, students often asked, "How do we know this is true? Our history books aren't telling the truth, why should we think this does?"

I explained the difference between primary and secondary sources and how a primary source — like a bill of sale or original list of slaves — was pretty solid evidence. To help ensure accuracy, the students decided that if we used secondary sources, we needed to find at least two different citations.

Bits and Pieces of Information

In the next several days the students, with my help, looked at various sources. We checked our school's children's books about presidents, our social studies textbook, a 1975 World Book Encyclopedia, and a CD-ROM encyclopedia. We found nothing about presidents as slave owners.

I had a hunch about which presidents owned slaves, based on what I knew in general about the presidents, but I wanted "proof" before we put a check in the "yes" box. And though my students wanted to add a third column — explaining how many slaves each slave-owning president had — that proved impossible. Even when we did find information about which presidents owned slaves, the numbers changed depending on how many slaves had been bought, sold, born, or died.

EIGHT COMPANIES EARN MORE THAN HALF THE WORLD'S POPULATION!

ONE-DOLLAR BILL GRAPHIC TEACHING SUGGESTIONS

Have students research data on the distribution of the world's wealth to see why Polyp, a graphic artist, created the image on the page at right.

Using such data, have students draw graphs and cartoons to demonstrate the inequality in the world's wealth.

Also do the simulation "Poverty and World Wealth," page 64.

In our research, most of the information dealt with presidential attitudes and policies toward slavery. It was difficult to find specific information on which presidents owned slaves. To help the investigation, I checked out a few books for them from our local university library.

Overall, our best resource was the Internet. The best sites required adult help to find and evaluate, and I became so engrossed in the project that I spent a considerable amount of time at home surfing the web. The "student-friendly" websites with information about presidents — such as the

know themselves." (Given more time, we might have explored this matter further, looking at who produces textbooks and why they might not include information about presidents' attitudes about racism and slavery.)

During our research, my students and I found bits and pieces of information about presidents and slavery. But we never found that one magic

POLYP (c) 2005 All rights reserved.

resource, be it book or website, that had the information readily available. Ultimately, though, we came up with credible data.

White House's gallery of presidents, www. whitehouse.gov/history/presidents — don't mention that Washington enslaved African Americans or that Jefferson did. Other popular sites with the same glaring lack of information are the Smithsonian Institution, http://educate.si.edu/president, and the National Museum of American History, www.americanhistory.si.edu/presidency.

As we did the research, I regularly asked, "Why do you think this doesn't mention that the president owned slaves?" Students' responses varied including "They're stupid," "They don't want us kids to know the truth," "They think we're too young to know," and "They don't

I'm a history buff, and I had thought I was on top of the question of presidents and slavery. I was quite amazed, and didn't hide my amazement from our action research team, when they discovered that two presidents who served after Lincoln — Andrew Johnson and Ulysses S. Grant — had been slave owners. While the students taped an extension on their chart, I explained that I was not totally surprised about Johnson because he had been a Southerner. But it was a shock that Grant had owned slaves. "He was the commander of the Union army in the Civil War," I explained. "When I first learned about the Civil War in elementary school, Grant and Lincoln were portrayed as saviors of the Union and freers of slaves."

Slave-Owning Presidents

When I told the entire class how Grant's slave-owning past had surprised me, Tanya, an African-American student, raised her hand and said, "That's nothing. Lincoln was a slave-owner, too."

I asked for her source of information and she said she had heard that Lincoln didn't like blacks. I thanked her for raising the point, and told the class that while it was commonly accepted by historians that Lincoln was not a slave-owner, his attitudes toward blacks and slavery were a source of much debate. I noted that just because a president didn't own slaves didn't mean that he supported freedom for slaves or equal treatment for people of different races.

I went into a bit of detail on Lincoln, in part to counter the all-too-common simplification that Lincoln unequivocally opposed slavery and supported freedom for blacks. I explained that while it's commonly believed that Lincoln freed enslaved Americans when he signed the Emancipation Proclamation, the document actually frees slaves only in states and regions under rebellion — it did not free slaves in any of the slaveholding states and regions that remained in the Union. In other words, Lincoln "freed" slaves everywhere he had no authority and withheld freedom everywhere he did. Earlier, in Lincoln's first inaugural address in March 1861, he promised slaveholders that he would support a constitutional amendment forever protecting slavery in the states where it then existed — if those states would only remain in the Union.

By the time we finished our research, the students had found that 10 of the first 18 presidents were slave owners: George Washington, Thomas Jefferson, James Madison, James Monroe, Andrew Jackson, John Tyler, James K. Polk, Zachary Taylor, Andrew Johnson, and Ulysses S. Grant.

Those that didn't own slaves were: John Adams, John Quincy Adams, Martin Van Buren, William Harrison, Millard Fillmore, Franklin Pierce, James Buchanan, and, despite Tanya's assertion, Abraham Lincoln.

The student researchers were excited to present their findings to their classmates and decided to do so as part of a math class. I made blank charts for each student in the class, and they filled in information provided by the action research team: the names of presidents, the dates of their years in office, the total number of years in office, and whether they had owned slaves. Our chart started with George Washington, who assumed office in 1789, and ended in 1877 when the last president who had owned slaves, Ulysses Grant, had left office.

> "[W]hile the word 'racism' does not appear, the subject of unfair treatment of people because of their race is addressed on page 467."
>
> — **The vice president of Harcourt School Publishers**

We then used the data to discuss this topic of presidents and slave-owning within the structure of ongoing math topics in my class: "What do the data tell us?" and "How can we construct new knowledge with the data?"

Students, for example, added up the total number of years in which the United States had a slave-owning president in office, and compared that total to the number of years in which there were non-slave-owning presidents in office. We figured out that in 69 percent of the years between 1789 and 1877, the United States had a president who had been a slave-owner.

One student observed that only slave-owning presidents served more than one term. "Why didn't they let presidents who didn't own slaves serve two terms?" another student pondered.

Using the data, the students made bar graphs and circle graphs to display the information. When they wrote written reflections on the math lesson, they connected math to content. One boy wrote: "I learned to convert fractions to percent so I know that 10/18 is the same as 55.5 percent. That's how many of the first 18 presidents owned slaves." Another girl observed, "I learned how to make pie charts and that so many more presidents owned slaves than presidents who didn't own slaves."

During a subsequent social studies lesson, the three students who had done most of the research explained their frustrations in getting information. "They hardly ever want to mention it [slaves owned by presidents]," explained one student. "We had to search and search."

Mini-Unit Objectives

Specific objectives for this mini-unit, such as reviewing the use of percentages, emerged as the lessons themselves unfolded. But its main purpose was to help students critically examine the actions of early leaders of the United States and become skeptical of textbooks and government websites as sources that present the entire picture. I figure that if kids start questioning the "official story" early on, they will be more open to alternative viewpoints later on. While discovering which presidents were slave owners is not an in-depth analysis, it pokes an important hole in the god-like mystique that surrounds the "founding fathers." If students learn how to be critical of the icons of American past, hopefully it will give them permission and tools to be critical of the elites of America today.

Besides uncovering some hard-to-find and uncomfortable historical truths, I also wanted to encourage my students to think about why these facts were so hard to find and to develop a healthy skepticism of official sources of information. I showed them two quotations about

> "As far as the laws of mathematics refer to reality, they are not certain, and as far as they are certain, they do not refer to reality."
>
> — **Albert Einstein,**
> *Sidelights on Relativity,* 1923

Thomas Jefferson. One was from a recently published, fifth-grade history textbook, *United States: Adventures in Time and Place* (New York: Macmillan/McGraw-Hill, 1998) which read: "Jefferson owned several slaves in his lifetime and lived in a slave-owning colony. Yet he often spoke out against slavery. 'Nothing is more certainly written in the book of fate than that these people are to be free.'" (page 314)

The other quotation was from James Loewen's *Lies My Teacher Told Me.* Loewen writes:

> Textbooks stress that Jefferson was a humane master, privately tormented by slavery and opposed to its expansion, not the type to destroy families by selling slaves. In truth, by 1820 Jefferson had become an ardent advocate of the expansion of slavery to the western territories. And he never let his ambivalence about slavery affect his private life. Jefferson was an average master who had his slaves whipped and sold into the Deep South as examples to induce other slaves to obey. By 1822, Jefferson owned 267 slaves. During his long life, of hundreds of different slaves he owned, he freed only three and five more at his death — all blood relatives of his. (page 140)

We talked about the different perspective each quote had toward Jefferson and toward what students should learn. My students' attention immediately turned to the set of spanking new history textbooks that had been delivered to our classroom that year as part of the districtwide social studies adoption. Some students assumed that our new textbook *United States* (Harcourt Brace, 2000) was equally as bad as the one I quoted from. One student suggested we just throw the books away. But I quickly pointed out they were expensive, and that we could learn from them even if they had problems and omissions.

I then explained what an omission was and suggested that we become "textbook detectives" and investigate what our new social studies text said about Jefferson and slavery. I reviewed how to use an index and divided all page references for Jefferson among small groups of students. The groups read the pages, noted any references to Jefferson owning slaves, and then reported back to the class. Not one group found a single reference.

Not surprisingly, the students were angry when they realized how the text omitted such important information. "They should tell the truth!" one student fumed.

No Mention of Racism

I wanted students to see that the textbook's omissions were not an anomaly, but part of a pattern of ignoring racism in the United States — in the past and in the present.

In the next lesson, I started by writing the word "racism" on the board. I asked the kids to look up "racism" in the index of their social studies book. Nothing. "Racial discrimination." Nothing.

"Our school should get a different book," one student suggested.

"Good idea," I said, "but it's not so easy." I told my students that I had served on a committee that had looked at the major textbooks published for fifth graders and that none of them had dealt with racism or slavery and presidents.

Students had a variety of responses:

"Let's throw them out."

"Let's use the Internet."

"Write a letter to the people who did the books."

I focused in on the letter-writing suggestion and reminded them that before we did so, we had to be certain that our criticisms were correct. The students then agreed that in small groups they would use the textbook's index and read what was said about all the first 18 presidents, just as we had done previously with Jefferson.

None of the groups found any mention of a president owning a slave.

Letters as Critique and Action

In subsequent days, some students wrote letters to the textbook publisher. Michelle, a white girl, was particularly detailed. She wrote: "I am 11 years old and I like to read and write. When I am reading I notice every little word and in your social studies book I realize that the word "racism" is not in your book. You're acting like it is a bad word for those kids who read it." She went on to criticize the book for not mentioning that any presidents had slaves: "I see that you do not mention that some of the presidents had slaves. But some of them did. Like George Washington had 317 slaves. So did Thomas Jefferson. He had 267 slaves." She continued: "If you want to teach children the truth, then you should write the truth." (Michelle's letter and some of the student-made charts were also printed in our school newspaper.)

We mailed off the letters, and moved on to new lessons. Weeks passed with no response and eventually the students stopped asking if the publishers had written back. Then one day a fancy-looking envelope appeared in my mailbox addressed to Michelle Williams. She excitedly opened the letter and read it to the class.

Harcourt School Publishers Vice-President Donald Lankiewicz had responded to Michelle at length. He wrote that "while the word 'racism' does not appear, the subject of unfair treatment of people because of their race is addressed on page 467." He also argued: "There are many facts about the presidents that are not included in the text simply because we do not have room for them all."

Michelle wrote back to Lankiewicz, thanking him but expressing disappointment. "In a history book you shouldn't have to wait till page 467 to learn about unfair treatment," she wrote. As to his claim that there wasn't room for all the facts about the presidents, Michelle responded: "Adding more pages is good for the kids because they should know the right things from the wrong. It is not like you are limited to certain amount of pages. ... All I ask you is that you write the word "racism" in the book and add some more pages in the book so you can put most of the truth about the presidents."

Michelle never received a reply.

Improving the Lesson

Michelle and the other students left fifth grade soon after the letter exchange. In the flurry of end-of-year activities, I didn't take as much time to process the project as I might have. Nor did I adequately explore with students the fact that most non-slave-owning presidents exhibited pro-slavery attitudes and promoted pro-slavery policies.

But the larger issue, which critical teachers struggle to address, is why textbook publishers and schools in general do such a poor job of helping students make sense of the difficult issues of race. We do students a disservice when we sanitize history and sweep uncomfortable truths under the rug. We leave them less prepared to deal with the difficult issues they will face in their personal, political, and social lives. Granted, these are extremely complicated issues that don't have a single correct response. But it's important to begin with a respect for the truth and for the capacity of people of all ages to expand their understanding of the past and the present, and to open their hearts and minds to an ever-broadening concept of social justice.

I believe my students learned a lot from their research on presidents and slaves — and clearly know more than most Americans about which of the first 18 presidents owned slaves. I'm also hopeful they learned the importance of looking critically at all sources of information.

I know one student, Tanya, did. On the last day of school she came up to me amid the congratulatory good-byes and said, "I still think Lincoln owned slaves."

"You are a smart girl but you are wrong about that one," I responded.

"We'll see," she said, "You didn't know Grant had slaves when the school year started! Why should I always believe what my teachers say?" □

Some of the students' names in this article have been changed.

A chart for students to record which presidents owned slaves appears on page 148. Copies of the correspondence and PDFs of additional handouts used in this lesson are available online at www.rethinkingschools.org/math.

Author's Note: About two years after I completed the research on slave-owning presidents with my students, folks at Wesleyan University put up a wonderful website, UnderstandingPrejudice.org. The page www.understandingprejudice.org/slavery includes extensive information on presidents who owned slaves. I learned from this website that three presidents not on my list also owned slaves: Martin Van Buren, William Henry Harrison, and James Buchanan. I was grateful for the additional information on this website, which opens up all sorts of new teaching possibilities.

PRESIDENTIAL FACTS

WHICH PRESIDENTS OWNED SLAVES AND FOR HOW LONG?

#	NAME OF PRESIDENT	YEARS IN OFFICE	TOTAL YEARS SERVED	OWNED SLAVES?	# OF SLAVES OWNED	YEARS SERVED BY SLAVE OWNER	YEARS SERVED BY NON-SLAVE OWNER
1	George Washington	1789–1797	8	Yes		8	
2							
3							
4							
5							
6							
7							
8							
9							
10							
11							
12							
13							
14							
15							
16							
17							
18							

LIBRARIES, BOOKS, AND BIAS

USING MATH TO RAISE ISSUES OF REPRESENTATION

BY BOB PETERSON

Classroom and school libraries offer many opportunities to apply math while raising issues of social justice and representation. For example: Who has access to how many books? Who is represented in those books?

Educational researcher Stephen Krashen argues that a key to improving student reading is access to large quantities of books. To have access to lots of books, schools need well-funded libraries staffed by qualified librarians. Others have argued that students should have access to books that reflect their own cultural heritage and those of many other cultures.

DO THE MATH

In small groups or as a whole class, students can investigate the following issues using ratios, percentages, graphs, and basic computation.

1.   What is the ratio of certified librarians to students in your school, school district, state, and nation? Make graphic representations of the data and then reflect on why there are differences, what impact these differences have, and what might be done to improve the situation.

2.   What is the ratio of books in your school library to the number of students in your school? Contact other schools or use the Internet to find similar data for other schools. Compare and contrast the data from urban, suburban, and rural school districts.

3.   Who is represented in the biographies in your school library? Count the number of biographies and autobiographies in your library. Calculate the percentage of books in your library that are in these categories. Make graphs showing the percentage of biographies that are about different groups of people, such as:

- Women/men.
- African Americans/Latinos/Asians/Native Americans/Arab Americans/European Americans.
- People from outside of the United States/people from the United States.
- Rich people/poor and working people.

4.   What do you think would be a "fair" representation in your school library of these various categories? What would have to be done to make sure there is an improvement in representation at your school?

In each of the projects described above, students could interview librarians, school district officials, and school board members regarding these questions. ∎

For more information on Stephen Krashen's work, visit www.sdkrashen.com or www.rethinkingschools.org/math.

THE MATHEMATICS OF LOTTERIES

GOING BEYOND PROBABILITY

BY LAURIE RUBEL

Many curricula point teachers and students toward using lotteries as a way to teach and learn probabilistic concepts. However, lotteries can be analyzed in sociopolitical context. This activity points students toward investigating critical questions like: Who typically plays the lottery? Who promotes lotteries? Who benefits from lottery revenues?

Studies show that people earning more than $100,000 annually are less likely to be pathological gamblers than people with lower incomes. Also, while the majority of gamblers are white, African Americans are disproportionately represented. Thus, poor people of color are, in general, most impacted by the mathematics of lotteries.

PROCEDURE

Groups of students can select one of their state lottery games and do three things:

1. Explain the game rules.

2. Provide a probability distribution of all possible outcomes.

3. Determine the expected value of the game for various jackpot amounts.

For example, the New York MegaMillions game is played by choosing five different integers from 1 to 52 as well as a MegaBall number, chosen from 1 to 52. The winning five numbers can be chosen in any of 52C5, or 2,598,960, ways. There are 135,145,920 different MegaMillions tickets,

MEGAMILLIONS

PRIZE	5 WINNING NUMBERS	MEGABALL	NUMBER OF TICKETS	PRIZE (ticket cost: $1)
1ST	Match 5	Match	1	Jackpot
2ND	Match 5	No match	51	$175,000–$1
3RD	Match 4	Match	235	$5,000–$1
4TH	Match 4	No match	11,985	$150–$1
5TH	Match 3	Match	10,810	$150–$1
6TH	Match 3	No match	162,150	$10–$1
7TH	Match 2	Match	551,310	$7–$1
8TH	Match 1	Match	891,825	$3–$1
9TH	No match	Match	1,533,939	$2–$1
NO PRIZE	No match	No match	131,983,614	Lose $1

but only one winner. Secondary prizes range from $175,000 to $2 for matching fewer than all of the numbers. Second prize winners match all five numbers but not the MegaBall number. So 51 of the 135,145,920 tickets win a second prize. Table 1 (page 150) shows the prize categories and the number of winning tickets. This shows that 3,162,306 tickets win, although about half win only $1. However, a staggering 131,983,614 are losing tickets!

QUESTIONS

The National Gambling Impact Study, commissioned by Congress in 1996, suggests that some state lotteries target poor communities for their advertising. Students can examine the realities in their communities:

1.   Where are the lottery retailers and where is the lottery advertising most prevalent?

2.   Who, in their communities, seems to be playing the lottery?

3.   Among those who play the lottery, how much do they spend each week?

FURTHER ANALYSIS

Once students consider these aspects of the state lottery, they can learn who benefits from lottery revenues. In New York, in the 2003–2004 fiscal year, the lottery directed 33 percent of its revenues ($1.9 billion) to state funds earmarked for education. New York City received about $703 million that year, earmarked toward education. But New York City education needs are far greater than those lottery revenues, so the lottery gets used as a kind of fiscal shell — extra money isn't given to education from lottery revenues; the lottery revenues enable the state to actually budget less for education. Students can analyze local changes in education budgets alongside rising lottery revenues to examine this issue.

Students can also uncover who benefits from lottery revenues, other than state departments of education, such as the lottery administration, lottery retailers, and game contractors. For example, in 2004, New York paid nearly $62 million in fees to its gaming contractor and additional money was spent on marketing and retailer commissions. Scientific Games Corp., a publicly traded company, is a lottery contractor who reported revenues of $560.9 million (!!) in 2003. The price of stock in this company has risen about $8 per share in the last year alone. Thus, an investment in the company which makes the lottery tickets and machines is astoundingly more sound than playing the lottery itself.

By their very nature, lottery games provide rich contexts for mathematical analysis. But distilling the mathematics of the games from their sociopolitical context, as is typically done, is a missed opportunity to help students use that mathematics to think critically about their own worlds. ∎

CHAPTER SEVENTEEN

The Transnational Capital Auction

A Game of Survival

BY BILL BIGELOW

Students are familiar with transnational corporations — Disney, Nike, Sony, Pepsi, Toyota. Corporations are entities with recognizable slogans and logos; they have public faces. Articles in the popular media describe the fortunes of this or that transnational corporation, the behavior of this or that CEO.

But transnational "capital" has no face. It has no jingles, no slogans, no logos. Capital is real, but invisible.

I wrote this simulation game, The Transnational Capital Auction, because I wanted students to grasp some aspects of capital as a force in today's world — to help them see capital as a kind of living being, one with certain needs and inclinations.

The game is a metaphor for the auction that capital holds to determine who in the world will make the most attractive bid for its "services." Students engage this dynamic from the standpoint of Third World elites and simulate a phenomenon that has been called the "race to the bottom," whereby these elites compete against one another to attract capital. The game's "punch line" is an examination of the social and ecological consequences of the auction.

Note: If any of what follows starts to feel complicated, don't worry. The game has been used by lots of teachers, with excellent success. It's more simple than it may appear — and it works. See my article "The Human Lives Behind the Labels" in *Rethinking Globalization* for a description of the activities that I've used before and after the game. (See Resources, page 170.)

Materials Needed

1. Several desirable candy bars — at least six.
2. Copies of the student handout, "Transnational Capital Auction: Instructions" (page 156) — one for each student in the class.
3. Copies of the "Transnational Capital Auction Credit Sheet" (page 157) — a minimum of seven, enough so that each country group can have at least one.
4. A minimum of 35 copies of the "Bids to Capital" slips (page 158) so that each group has one per round (seven groups, five auction rounds.)

Suggested Procedure

1. Before the activity, create the auction scoring chart on the board or on an overhead transparency (see page 155).

2. If you have at least 21 students in your class, divide them into seven groups. Ask the groups to form around the classroom, as far away from one another as possible.

3. Distribute copies of "Transnational Capital Auction: Instructions," "Transnational Capital Auction Credit Sheet," and the "Bids to Capital" slips. Read aloud with students "Transnational Capital Auction: Instructions." It should be obvious, but emphasize the distinction between "Friendly to Capital" credits and game points.

Answer any questions students might have. Review the "Transnational Capital Auction Credit Sheet." Point out that a group earns more credits the friendlier it is to capital. Show them the candy bars and announce that the three groups with the most game points will win all the candy bars.

4. Begin the first round: Tell students to make their bids in each of the categories and to total up their Friendly to Capital credits on the "Bids to Capital" slips. Note that this is really the hardest round because students don't have any way of knowing what the other groups are bidding.

I play "Capital" in the game and wander the classroom as the small groups decide on their bids, urging them to lower the minimum wage, taxes on corporate profits, and the like: "Come on, show that you really want me!"

5. After each group has submitted its bid, write these up on the board or overhead. Award the first game points based on the results — again, 100 game points for the third highest number of Friendly to Capital credits, 50 for the second highest and 25 for the first. (This is explained in the instructions handout.) After this first round has been played, and the points are posted, I scoff at the losers and urge them to get with the program and start making some bids that will attract Capital.

From this point on, for better or worse, the competition to "win," or for candy bars, takes over, and students continue to "race to the bottom" of conditions for their respective countries. Sometimes they even realize what they're doing as they decide on their bids. "This is like 'The Price is Right,' oppression style," I overheard one of my students say one year.

6. After each round of bids, continue to post the Friendly to Capital credit scores and

Quality, Industriousness and Reliability Is What El Salvador Offers You!

Rosa Martinez produces apparel for U.S. markets on her sewing machine in El Salvador. You can hire her for 57-cents an hour*.

Rosa is more than just colorful. She and her co-workers are known for their industriousness, reliability and quick learning. They make El Salvador one of the best buys in the C.B.I. In addition, El Salvador has excellent road and sea transportation (including Central America's most modern airport) . . . **and there are no** quotas.

Find out more about **sourcing** in El Salvador. Contact **FUSADES**, the private, non-profit and non-partisan organization promoting social and economic development in El Salvador. Miami telephone: **305/381-8940.**

* - Does not include fringe benefits.

1990

Quality, Industriousness and Reliability Is What El Salvador Offers You!

Rosa Martinez produces apparel for U.S. markets on her sewing machine in El Salvador. You can hire her for 33-cents an hour*.

Rosa is more than just colorful. She and her co-workers are known for their industriousness, reliablity and quick learning.
They make El Salvador one of the best buys.
In addition, El Salvador has excellent road and sea transportation (including Central America's most modern airport) . . . **and there are no** quotas.

Find out more about **sourcing** in El Salvador. Contact **FUSADES**, the private, non-profit and non-partisan organization promoting social and economic development in El Salvador. Miami telephone: **305/529-2233,** Fax: **305/529-9449.**

*Does not include fringe benefits/exact amounts may vary slightly depending on daily exchange rates.

1991

THESE ADS APPEARED IN TRADE MAGA-ZINES FOR CLOTHING MANUFACTURERS. NOTE THAT THE HOURLY WAGE DROPPED FROM 57 CENTS AN HOUR IN THE FIRST AD TO 33 CENTS IN THE SECOND.

award game points for that round. Keep a running total of each country team's game points. As Capital, I continue to urge students lower and lower: "Team five, you think I'm going to come to your country if your tax rate is 30 percent? Come on, next round let's get that way down."

7. For the fifth and final round, ask each student to write down their group's last bid, separate from the "Bids to Capital" slips — not just the number of credits, but the actual minimum wage, the child labor laws, etc. For their homework writing assignment, each student needs to know the specific social and environmental conditions created by the auction, in their pretend country.

Award candy bars to the "winners." Distribute the homework assignment, "Transnational Capital Auction Follow-up" (page 159).

8. The next day, we discuss the homework questions on the "Transnational Capital Auction Follow-up" student handout. Some additional questions to raise:

- What does it mean to have no environmental laws in a country? What might Capital do?
- What would be the social effects of such low wages? (You might list these on the board or overhead.) How would families be able to survive? How could a family supplement its income? You might point out that even if a country did have child labor laws, the low wages for adults would put pressure on a family to send its children to work. If students don't point it out, you might also note the relationship between the "race to the bottom" and increased immigration. Of course, there are other factors leading to immigration, such as the dramatic rise in cash crop agriculture which throws peasants off the land. (See, for example, the video *The Business of Hunger*, listed in Resources, page 173.) But it's vital that students begin to see the interconnectedness of global issues.
- Ask students to look at the El Salvador ads from 1990 and 1991 (at left). How are these ads examples of the "race to the bottom"? Who are the ads trying to appeal to?
- With homework question 2 on the handout, I want students to begin to think critically about some of the ideas that are presented as obvious and absolute in much of the media: Is corporate investment always a good thing? You might make two columns on the board or overhead and ask students to list some of the possible benefits of investment and some of the harmful effects of investment and how it is attracted.

Encourage students to ground their answers in their own experiences with the Transnational Capital Auction, as my student Sam did in the following part of his answer to question 2: "Poor countries may need investment, but it is not necessarily a good thing when transnational companies invest capital there. I come to this conclusion by the simple attraction of Capital. Capital is attracted to a place that has less restrictions, on everything. In order to make our country more 'Capital friendly' we were willing to sell out the lives of individuals more and more, step by step. The worse the conditions got, the more Capital was interested."

- Why did you keep driving down conditions in your country? Why didn't you get together and refuse to bid each other down?
- Who benefits and who doesn't benefit from the "race to the bottom" we simulated in class? How could people in various countries get together to stop attacks on their social and environmental conditions?
- What could we do in this country to respond to the global "race to the bottom"?

9. One option is to give students an opportunity to play the Transnational Capital Auction all over again, except this time they could represent labor and environmental activists instead of a country's elite.

It's important that, if possible, students not be left with the sense that the downward leveling they experience in the auction is inexorable. There are things that people can do, and are doing. You might, for example, watch the video *Global Village or Global Pillage* (see Resources, page 173). Students need to see "big pictures" in order to understand seemingly disconnected events in countries around the world. But it's vital that they not feel defeated by this awareness. □

TRANSNATIONAL CAPITAL AUCTION

SCORING CHART

COUNTRY	ROUND 1		ROUND 2		ROUND 3		ROUND 4		ROUND 5	
	friendly to capital credits	game points	friendly to capital credits	game points	friendly to capital credits	game points	friendly to capital credits	game points	friendly to capital credits	game points
1										
2										
3										
4										
5										
6										
7										

TRANSNATIONAL CAPITAL AUCTION

INSTRUCTIONS

You are leaders of a poor country. Each of your countries was either colonized by European countries or dominated by them economically and militarily. You need to attract foreign investment (capital) from transnational corporations for many different reasons.

Of course, not all of your people are poor. Many, including a number of you, are quite wealthy. But your wealth depends largely on making deals with corporations that come to your country. You get various kickbacks, bribes, jobs for members of your families, etc. Some of this is legal; some not. But in order to stay in power you also need to provide jobs for your people, and the owners of capital (companies like Nike, Disney, Coca-Cola, etc.) are the ones who provide thousands of jobs in their factories.

THE CHALLENGE

The more jobs you can bring into your country, the more legitimacy you have in the eyes of your people. And your government collects taxes from these companies, which help keep your government working and also help you pay back loans to the International Monetary Fund and other foreign-owned banks. The bottom line is this: You badly need these companies to invest capital in your country.

But here's the problem: You must compete with other poor countries that also need capital. Corporations are not stupid, so they let you know that if you want their investment, you must compete with other countries by:

• Keeping workers' wages low.
• Having few laws to regulate conditions of work (overtime, breaks, health and safety conditions, age of workers), or not enforcing the laws that are on the books.
• Having weak environmental laws.
• Making sure that workers can't organize unions; maintaining low taxes on corporate profits, etc.

Basically, companies hold an auction for their investments. The countries who offer the companies the most "freedom" are the ones who get the investment.

THE GAME

The goal is to win the game by ending up with the most game points after five auction rounds. Each country team's goal is to "win" by attracting capital. The team that bids the third highest number of "Friendly to Capital" credits in a round is awarded 100 game points; the team with the second highest number of Capital credits is awarded 50 game points; and the team with the highest number of Capital credits is awarded 25 game points. The other teams get no points for the round. The auction is "silent" — you don't know until the end of each round who has bid what.

Again, Capital will go where the people are "friendliest" to it. However, the "friendlier" you are to Capital, the angrier it may make your own people. For example, Capital wants workers to work for very little and to not worry about environmental laws. But that could start demonstrations or even rebellions, which would not be good for Capital or for you as leaders of your country. That's why the team bidding the highest number of Capital credits does not get the highest number of game points.

THE LAST RULE

Your team may be the highest (Capital credit) bidder twice and not be penalized. But for each time you are highest bidder more than twice, you lose ten game points — 10 the first time, 20 the second, etc. This is a "rebellion penalty."

Good luck.

TRANSNATIONAL CAPITAL AUCTION **CREDIT SHEET**

minimum wage/hour	credits
$5.00	0
$4.75	10
$4.50	15
$4.25	20
$4.00	25
$3.75	30
$3.50	33
$3.25	37
$3.00	40
$2.75	43
$2.50	46
$2.25	49
$2.00	52
$1.75	55
$1.50	58
$1.25	61
$1.00	64
$0.85	67
$0.75	70
$0.65	73
$0.55	76
$0.45	79
$0.35	82
$0.30	85
$0.25	88
$0.20	91
$0.15	94
$0.10	97
$0.05	100

child labor	credits
below age 16 is illegal/enforced	0
below age 16 is illegal/weakly enforced	15
below age 16 is illegal/not enforced	30
below age 14 is illegal/enforced	50
below age 14 is illegal/weakly enforced	70
below age 14 is illegal/not enforced	85
no child labor laws	100

environmental laws	credits
strict laws/enforced	0
strict laws/weakly enforced	15
strict laws/not often enforced	30
some laws/enforced	50
some laws/weakly enforced	70
some laws/not often enforced	85
almost no environmental laws	100

worker organizing	credits
unions fully legal/allowed to organize	0
unions fully legal/some restrictions on right to strike	15
only government approved unions legal/ some restrictions on right to strike	30
only government organized unions allowed	45
unions banned/no right to strike	60
unions banned/no right to strike/military stationed in factories	85
unions banned/no right to strike/military stationed in factories/suspected union organizers jailed/military used against strikes	100

taxation of corporate profits

rate	credits	rate	credits	rate	credits	rate	credits
75%	0	55%	20	35%	40	15%	75
70%	5	50%	25	30%	50	10%	80
65%	10	45%	30	25%	60	5%	90
60%	15	40%	35	20%	70	0% (no tax)	100

TRANSNATIONAL CAPITAL AUCTION

BIDS TO CAPITAL

country #_____

round #_____

minimum wage credits > > > > _____

child labor credits > > > > > > > _____

worker organizing credits> > > _____

taxation rate credits> > > > > > > _____

environmental laws credits > _____

TOTAL CREDITS THIS ROUND _____

TRANSNATIONAL CAPITAL AUCTION

BIDS TO CAPITAL

country #_____

round #_____

minimum wage credits > > > > _____

child labor credits > > > > > > > _____

worker organizing credits> > > _____

taxation rate credits> > > > > > > _____

environmental laws credits > _____

TOTAL CREDITS THIS ROUND _____

TRANSNATIONAL CAPITAL AUCTION

BIDS TO CAPITAL

country #_____

round #_____

minimum wage credits > > > > _____

child labor credits > > > > > > > _____

worker organizing credits> > > _____

taxation rate credits> > > > > > > _____

environmental laws credits > _____

TOTAL CREDITS THIS ROUND _____

TRANSNATIONAL CAPITAL AUCTION

BIDS TO CAPITAL

country #_____

round #_____

minimum wage credits > > > > _____

child labor credits > > > > > > > _____

worker organizing credits> > > _____

taxation rate credits> > > > > > > _____

environmental laws credits > _____

TOTAL CREDITS THIS ROUND _____

TRANSNATIONAL CAPITAL AUCTION

FOLLOW-UP

Complete these on a separate sheet of paper.

1.   Look over your auction "bids" for the fifth and final round of the Transnational Capital Auction — on minimum wage, child labor, worker organizing, taxation rates, and environmental laws. If Capital were to accept your "bid" and come to your country, what would be the real human and environmental consequences there? Answer this question in detail.

2.   Based on your experience with the auction, agree and/or disagree with the following statement and back up your answer with evidence: "Poor countries need investment, so it's a good thing when transnational companies invest there."

3.   The global process that we simulated in class is sometimes called "downward leveling" or the "race to the bottom." What, if anything, could people in poor countries do to stop this race to the bottom?

4.   One company used to manufacture all its products in the United States, paying wages that averaged (with benefits) around $16 an hour. The investment director for this company now travels every month to places like Indonesia, El Salvador, and Nicaragua looking for sites to produce his company's products. He says that he would prefer to keep all production in the United States. Based on this simulation and what you know, why do you think this person's company feels forced to send production to countries that have a lot of Friendly to Capital points?

5.   What impact does the race to the bottom have on workers in this country? In what ways might it affect your lives? In answering the question, you might think about the three quotes below.

"It is not that foreigners are stealing our jobs,
it is that we are all facing one another's competition."
— **William Baumol,** economist, Princeton University

"Downward leveling is like a cancer that is destroying its host
organism — the earth and its people."
— **Jeremy Brecher and Tim Costello,** *Global Village or Global Pillage*

"Globalization has depressed the wage growth of low-wage workers [in the United States]. It's been a reason for the increasing
wage gap between high-wage and low-wage workers."
— **Laura Tyson,** former chair, U.S. Council of Economic Advisers

SWEATSHOP MATH

TAKING A CLOSER LOOK AT WHERE KIDS' "STUFF" COMES FROM

BY BOB PETERSON

Students are intrigued about their "stuff" and where their stuff comes from. It's a great entry point for teachers and students to explore various issues, including sweatshops, child labor, and consumerism. In *Rethinking Globalization* (see Resources, page 170) there are many background readings, lesson ideas, and resources to help teachers engage students in thinking deeply about these topics. In addition, *Rethinking Globalization* lists many videos on these subjects that also include some excellent math.

DO THE MATH

Numerous math opportunities exist when students explore these issues. Among them:

• Tabulate, figure the percentage, and graph which countries students' clothes, bookbags, and shoes come from. Locate all the countries on a wall map using small Post-it notes.

• Using data from *Rethinking Globalization* or from the websites listed below, find the daily, weekly and annual wages of people making common items that people in this country buy. Compare those wages with wages in other countries.

For example: Workers in the Evergreen Factory in the Rio Blanco Industrial Park sew McKids children's clothing, commonly sold at Wal-Mart stores, and make 43 cents an hour. If a worker works a 14-hour shift, how much does he or she make in one day? How much would he or she make in six days? Assuming a six-day work week, with no unpaid sick days or vacation, how much would he or she earn in a year?

QUESTIONS FOR DEEPER THINKING

• What additional data is necessary to determine how difficult it would be to live on certain wages in a country? How could you find that data?

- Why do companies move their operations to countries where the wages are low?

- When companies move jobs to lower-wage countries, what are the consequences for the workers and communities where the companies originally were located?

- What movements exist to improve the wages and working conditions of people working under these conditions? ∎

Excellent websites for collecting data and background information include:

Rethinking Globalization, www.rethinkingschools.org/rg

National Labor Committee, www.nlcnet.org

Campaign for Labor Rights, www.campaignforlaborrights.org

Clean Clothes Campaign, www.cleanclothes.org

Free the Children, www.freethechildren.org

Global Exchange, www.globalexchange.org

International Labor Organization, www.ilo.org

Sweat Shop Watch, www.sweatshopwatch.org

"WORK FASTER!"

FRED WRIGHT

CARTOON TEACHING SUGGESTIONS

Ask students: "What does the cartoonist suggest is the source of wealth?" and "How would one determine a 'fair wage' given various costs of making something?" "What percentage of the cost of the product does the worker get?"

Have students research the average CEO salary and compare it to the average production worker salary. Assume, for the mooment, that (a) the CEO gets the full amount that the worker does not get (in this case, $75 to the worker's $25), and (b) the product still costs $100.

DO THE MATH

Use the data on average CEO and production worker salaries to figure out their relative shares of the $100. Is the cartoon accurate? If not, how should it be adjusted?

Show students the segment of the video *Zoned for Slavery* by the National Labor Committee about the costs of making a Gap shirt in El Salvador. Have students redo the math in the above cartoon and discuss what it means for the people who work.

See "The Organic Goodie Simulation" in *Rethinking Our Classrooms, Vol. 1* (see Resources, page 163) for a great classroom activity that deals with this issue.

PART 4

Resources
for Rethinking
Mathematics

Websites

The Algebra Project

www.algebra.org

Started by civil rights activist Bob Moses, The Algebra Project works for access to higher level math classes for all people, particularly African Americans.

The Benjamin Banneker Association, Inc.

www.bannekermath.org

A nonprofit organization of individuals and groups concerned about the mathematics education of African-American children.

CorpWatch

www.corpwatch.org

Excellent collection of background articles (and cartoons) on a range of important issues: tobacco, sweatshops, pharmaceuticals, trade agreements, and more.

Dollars and Sense

www.dollarsandsense.org

Excellent quarterly magazine on economic justice. Check out their books as well. Especially useful for high school level math resources.

Ethnomathematics Digital Library

www.ethnomath.org

A collection of over 700 articles, many of them scholarly, on critical and ethnomathematics by scholars from around the world.

Fairness & Accuracy In Reporting

www.fair.org

A national media watch group that has offered well-documented criticism of media bias and censorship since 1986. FAIR publishes the excellent, classroom-friendly Extra! which is a vital source to get students thinking critically about media coverage of world events.

Global Exchange

www.globalexchange.org

An organization dedicated to promoting environmental, political, and social justice around the world. In the late 1990s, Global Exchange was perhaps the most important organization drawing attention to Nike's sweatshop abuses. The website details the group's many activities and resources.

The Harper's Index

www.harpers.org/HarpersIndex.html

A monthly statistical snapshot of the world's economic, political, and cultural climate. An excellent source of data for various types of problem solving.

Jobs With Justice

www.jwj.org

A national campaign, with local affiliates, to organize support for workers' rights struggles. Their Student Labor Action Project supports student activism around issues of workers' rights as well as social and economic justice.

National Council of Teachers of Mathematics

www.nctm.org

The principal professional organization of mathematics teachers in the United States. Offers various journals, conferences, and publications.

National Labor Committee

www.nlcnet.org

An organization that has led the fight against sweatshops in many countries. Articles include those on specific companies and countries. Specific data on wages, hours, and working conditions.

National Priorities Project

www.natprior.org

An excellent site for articles and facts on a variety of national issues such as the federal budget, the war in Iraq, prescription drugs, the environment, and others.

Orilla's Project: Connecting Math to Our Lives

www.orillas.org/math

An international, web-based project that connects math teachers around the world who are using math as a tool for social justice.

Project Seed

www.projectseed.org

A multifaceted mathematics education program for grades 3–8 that has been around for over 40 years. It uses a Socratic method to teach advanced mathematics to urban youth.

Radical Math

www.radicalmath.org

An excellent site developed by math teachers who have assembled hundreds of lessons, charts, graphs, articles, and data sets to help teachers integrate issues of social justice into math classes.

Rethinking Schools

www.rethinkingschools.org

Publisher of the book *Rethinking Mathematics*, the quarterly journal *Rethinking Schools,* and over a dozen other books. A specific page on the website, www.rethinkingschools/math, provides additional math teaching ideas to complement this book.

SACNAS (Society for the Advancement of Chicanos and Native Americans in Science

www.sacnas.org

The mission of SACNAS is to encourage Chicano/Latino and Native American students to pursue graduate education and obtain the advanced degrees necessary for science research, leadership, and teaching careers at all levels.

Teachers for Social Justice (Chicago)

www.teachersforjustice.org

Website has a calendar of activities and teaching lessons, including several on mathematics.

Teaching for Change

www.teachingforchange.org

The Teaching for Change online catalog is a great source for books, videos, and posters for teachers wanting to bring social justice into the classroom.

TODOS: Mathematics for All

www.todos-math.org

Advocates for an equitable and high-quality mathematics education for all students, in particular Latino/Hispanic students, by advancing the professional growth and equity awareness of educators.

United for a Fair Economy

www.faireconomy.org

An excellent resource for up-to-date data on current social/economic issues. Check out the free "workshops" on Fair Taxes for All, War and the Economy, FTAA for Beginners, The Growing Economic Divide, and others.

UNICEF (United Nations Children's Fund)

www.unicef.org

UNICEF produces educational materials and distributes funds to children's programs throughout the world. Their annual report on The State of the World's Children provides useful statistics.

Math Curriculum and Pedagogy

Education Development Center, Inc., **The Language of Numbers** (Heinemann, 1993). Twenty-five lessons for students in which they invent, explore, and compare number systems by using the Chinese abacus, the ancient Egyptian number system, and numbers in different languages. Grades 6 and 7.

Marilyn Frankenstein, **Relearning Mathematics: A Different Third R — Radical Maths** (Free Association Books, 1989). A comprehensive description of how to teach math from a social justice perspective, including chapters on fractions, decimals, percentages, and measurement.

Kevin J. Mistrik and Robert C. Thul, **Math for a Change** (Mathematics Teachers Association, 1993). Contains 41 situations of injustice that need mathematics to be fully understood. Problems are accompanied by teacher notes and include contexts such as welfare, AIDS, and the savings-and-loans scandals. Grades 7–12.

Kevin J. Mistrik and Robert C.Thul, eds., **Math for a World that Rocks** (Mathematics Teachers Association, 1997). Written and illustrated by students based on "Math for a Change." Covers 23 social justice projects with topics such as rain forests, strip mining, and DWUI.

Robert P. Moses and Charles E. Cobb, **Radical Equations: Math Literacy and Civil Rights** (Beacon Press, 2001). The authors argue for mathematical literacy as the new "civil right," and they advocate organizing people to demand educational access to powerful mathematics. Moses also tells the story of the voter registration movement in Mississippi during the 1960s which, he argues, holds crucial lessons for today.

National Council of Teachers of Mathematics, **Multicultural and Gender Equity in the Mathematics Classroom: The Gift of Diversity** (NCTM, 1997). The 1997 NCTM Yearbook. Aimed at classroom teachers, this volume has five sections: issues and perspectives; classroom cultures; curriculum, instruction, and assessment; professional development; and future directions. Mainly short articles, it is generally useful.

Leslie J. Nielson and Michael de Villiers, **Is Democracy Fair? The Mathematics of Voting and Apportionment** (Key Curriculum Press, 1997). This interesting book grows out of the new South Africa and its freedom from apartheid. In 1994 de Villiers wrote an initial volume (1) to point out that different voting schemes were both mathematical and political choices; (2) to educate people about the different schemes being considered for the new constitution; and (3) to challenge the stereotype that mathematics is only of value to the "hard" sciences. It has excellent teacher notes, is classroom ready, covers many different voting methods, and is very accessible.

Sharan-Jeet Shan and Peter Bailey, **Multiple Factors: Classroom Mathematics for Equality and Justice** (Trentham Books, 1991). A book from the U.K. about developing anti-racist and pro-justice mathematics teaching. The authors counter Eurocentric approaches to mathematics, examine the history of mathematics, and provide fresh approaches and plenty of examples for teachers, set in explicitly sociopolitical contexts.

Terry Vatter, **Civic Mathematics: Fundamentals in the Context of Social Issues** (Teacher Idea Press, 1996). A year's worth of middle school mathematics activities using social issues. The material covers the range of basic middle school math areas and is separated into four quarters: Issues of Race and Gender, Poverty and Wealth, The Environment, and Teen Issues. Each section has a range of activities that teachers can adapt for their particular context.

David Whittin and Sandra Wilde, **Read Any Good Math Lately?** Children's Books for Mathematical Learning, K–6 (Heinemann, 1992).
Teaching ideas and bibliographic information useful for integrating literature and mathematics instruction. It does not have a social justice orientation but is helpful nonetheless. Also see *It's the Story That Counts,* by the same authors.

Claudia Zaslavsky, **Africa Counts: Number and Pattern in African Cultures**, 3rd ed. (Lawrence Hill Books, 1999).
Zaslavsky's classic book on the mathematics of Africa covers how numbers and patterns are integral parts of daily and scientific life all over Africa. Includes a thorough debunking of Eurocentric biases and provides rich examples of mathematical recreations, patterns, shapes, counting and numeration systems, geometry, and art — from all around the continent.

Claudia Zaslavsky, **The Multicultural Math Classroom: Bringing in the World** (Heinemann, 1996).
A multicultural perspective for elementary and middle school teachers, including teaching ideas on number systems, counting, calculating, geometry, and data analysis.

Claudia Zaslavsky, **More Math Games and Activities from Around the World** (Chicago Review Press, 2003).
A compilation of 70 games, puzzles, and projects from around the world that help students become interested in and learn math while giving them a greater appreciation for the role of math in many countries. Also see *Math Games and Activities from Around the World.*

Math Books with Theoretical/ Academic Perspectives

Leone Burton, **Which Way Social Justice in Mathematics Education?** (Praeger Publishers, 2003).
Provides an international, research-based perspective on the impact of gender, race, ethnicity, culture, and social class in student mathematics learning and achievement. Although the main focus is on gender issues, the ideas touch broader sociocultural issues in mathematics education.

Gilbert Cuevas and Mark Driscoll, eds., **Reaching All Students with Mathematics** (NCTM, 1993).
This NCTM publication contains a lot of information on various programs oriented toward academic achievement and access for students of color and female students. Sections include: A Global View, Changing What Students Learn, Changing How Teachers Teach, Changing How Students Learn.

Eric Gutstein, **Reading and Writing the World with Mathematics: Toward a Pedagogy for Social Justice** (Routledge, 2006).
A theoretical and practical book about teaching and learning mathematics for social justice, based on a four-year, teacher-research study of the author's own classroom.

Carol Malloy and Laura Brader-Araje, **Challenges in the Mathematics Education of African American Children: Proceedings of the Benjamin Banneker Association Leadership Conference** (NCTM, 1998).
The conference proceedings, keynote address, and research reports from a Banneker

Association conference.

Danny Martin, **Mathematics Success and Failure Among African-American Youth** (Erlbaum Associates, 2000).
An excellent study of African-American mathematics students in two middle schools, combining sociocultural and cognitive approaches. The author examines issues of student identity and personal agency, community and cultural forces, history, and social context to develop an explanatory theory for why some students succeed while others do not.

David Nelson, George Gheverghese Joseph, and Julian Williams, **Multicultural Mathematics: Teaching Mathematics from a Global Perspective** (Oxford University Press, 1993).
This excellent resource from the U.K. presents a rationale for teaching mathematics with a multicultural and global perspective and provides many examples that teachers can learn from.

Walter G. Secada, series ed., **Changing the Faces of Mathematics** (NCTM, 1999, 2000).
This series covers a wide range of topics, approaches, and programs. The six volumes are: Perspectives on Latinos, Perspectives on Asian Americans and Pacific Islanders, Perspectives on African Americans, Perspectives on Multiculturalism and Gender Equity, Perspectives on Gender, and Perspectives on Indigenous People of North America. Excellent resources.

Walter G. Secada, Elizabeth Fennema, and Lisa B. Adajian, **New Directions for Equity in Mathematics Education** (Cambridge University Press, 1995).
A collection of articles about mathematics education reform and equity. Themes discussed include social class, race and ethnicity, culture, gender, and language. The articles also deal with teacher deskilling, classroom dynamics, and issues of power.

Ole Skovsmose, **Towards a Philosophy of Critical Mathematics Education** (Kluwer Academic Publishers, 1994).
Don't ask the price of a Kluwer book: Only libraries can afford them. But this is the book that lays out a comprehensive and philosophical framework for critical mathematics education. Skovsmose is Danish and writes from that perspective, but the ideas are applicable across the globe.

Valerie Walkerdine, **Counting Girls Out: Girls and Mathematics**, new ed. (Falmer Press, 1998).
In her classic post-structuralist account of girls in mathematics, Walkerdine argues that dominant ideologies (like patriarchy) play a key role in how we see girls (and boys) in school mathematics. She makes the point that the problem of girls' underachievement in mathematics is not located within girls, but rather in the sociopolitical context of society itself. An important theoretical work that helps us think more deeply about the relationships of gender, inequality, and ideology.

Curriculum Guides/Resources That Include Math

Ann and Johnny Baker, **Counting on Small Planets: Activities for Environmental Mathematics** (Heinemann, 1991).
A concise description of how to blend math and environmental education around issues of waste, noise, water, erosion, and deforestation.

Maurice Bazin and Modesto Tamez, **Math and Science Across Cultures** (New Press, 2002).
A unique collection of activities and lessons on how math is integrated into work and play in different cultures. Lessons include Madagascar Solitaire, The Inca Counting System, and Math in Ancient Egypt.

Bill Bigelow and Norm Diamond, **The Power in Our Hands: A Curriculum on the History of Work and Workers in the United State**s (Monthly Review, 1988).
Awidely used curriculum on labor history. Role plays, simulations, first-person readings, and writing activities help students explore issues of work and social change.

Bill Bigelow and Bob Peterson, **Rethinking Globalization: Teaching for Justice in an Unjust World** (Rethinking Schools, 2002).
An excellent resource for teaching strategies that help students make sense of an increasingly complicated and scary world. Includes role plays, interviews, poetry, stories, background readings, and hands-on teaching tools. Math lessons include Sweatshop Math.

Bill Bigelow, Linda Christensen, Stan Karp, Barbara Miner, and Bob Peterson, eds., **Rethinking Our Classrooms, Vol. 1: Teaching for Equity and Justic**e (Rethinking Schools, 1994).
A collection of lessons, reflections, poems, and resources for social justice teaching.

Bill Bigelow, Brenda Harvey, Stan Karp, and Larry Miller, eds., **Rethinking Our Classrooms, Vol. 2: Teaching for Equity and Justice** (Rethinking Schools, 2001).
Extends and deepens many of the themes introduced in Rethinking Our Classrooms, Vol. 1, which has sold more than 125,000 copies. Practical, from-the-classroom stories from teachers about how they teach for social justice.

Magi Black, **The No-Nonsense Guide to Water** (Verso Press and the New Internationalist, 2004).
Like other books in the No-Nonsense series, this informative handbook is packed with facts, charts, and essays that help teachers and students understand how water has become one of the most important resources in the world and how inequality in access to water is a global crisis. An excellent resource for science and math teachers.

Deborah Menkart, Enid Lee, Margo Okazawa-Rey, eds., **Beyond Heroes and Holidays** (NECA, 1998). A compilation of teaching and staff development activities that emphasize anti-racist, social justice approaches including ideas on teaching of math.

Tamara Sober Giecek, **Teaching Economics as If People Mattered** (United for a Fair Economy, 2000).
A teacher-friendly guide, full of reproducible graphics and specific teacher lessons on a range of important economic issues including income distribution, wages and salaries of workers and CEOs, globalization and more. High school and above, although some of the charts and activities could be modified for middle school and upper elementary.

Fred Gross, Patrick Morton, and Rachel Poliner, **The Power of Numbers: A Teacher's Guide to Mathematics in a Social Studies Context** (Educators for Social Responsibility, 1993). Though not explicitly social justice oriented, this book contains some interesting lessons on polling, graphing, and using census figures. Stehpanie Kempf, **Finding Solutions to Hunger: Kids Can Make a Difference** (World Hunger Year, 1997).

A source on hunger for middle and upper school teachers that draws in issues of social studies, math, and politics.

Anita Roddick with Brooke Shelby Biggs, **Troubled Water: Saints, Sinners, Truths and Lies about the Global Water Crisis** (Anita Roddick Books, 2004).

This amazing book of eclectic essays, facts, quotations, drawings, and diagrams makes readers look at the pending water crisis from a variety of viewpoints. Excellent for science, social studies, and math. All ages.

Jeremy Seabrook, **The No-Nonsense Guide to World Poverty** (Verso Press and the New Internationalist, 2003).

A handbook packed with facts, vignettes, and essays on the roots, scope, and impact of world poverty. An excellent resource for high school social studies and math teachers.

Additional Books Useful for Mathematics Teaching

Dollars and Sense and United for a Fair Economy, eds., **The Wealth Inequality Reader** (Dollars and Sense, 2004).

A collection of 25 articles that explore wealth inequality, its causes and consequences, and strategies for change. The graphics alone in the opening overview make the book worthwhile for any high school social studies, math, or economics teacher.

Nancy Folbre, **The New Field Guide to the U.S. Economy** (New Press, 1995).

An excellent guide to economic life in the United States. Very useful data, graphics, and graphs on a wide range of social issues. Good for upper elementary through college.

John Allen Paulos, **A Mathematician Reads the Newspaper** (Anchor, 1995).

A somewhat cynical look at the mathematical illiteracy of people in the United States, but not necessarily a critical look. Still, a very useful resource for teachers who want to think carefully about promoting critical numeracy.

Dan Smith and Ane Braein, **Penguin State of the World Atlas**, 7th ed. (Penguin Books, 2003).

Statistics are given shape and meaning through the visual analysis of data in full-color maps and graphics. Topics include: war, globalization, population growth, human rights, children's rights, health, and more.

Dan Smith, **The Penguin Atlas of War and Peace** (Penguin Books, 2003).

Packed with data analysis and essential political and historical background. Covers the causes of war, the costs of war, and trends in peace, from Afghanistan to Northern Ireland to Kosovo to the Middle East.

Ward L. Kaiser and Denis Wood, **Seeing Through Maps: The Power of Images to Shape Our World View** (ODT, Inc., 2001). An excellent resource for understanding how maps are created and some of the mathematical issues involved. This book helps us see how maps shape our view of ourselves and others, and examines issues in various world maps (e.g., the Peters and Mercator projections).

Children's Books

Diana Cohn, **¡Sí Se Puede!/Yes, We Can!** illus. Francisco Delgado (Cinco Puntos Press, 2004). Narrated by a child who tells the story of her mother, a janitor in a Los Angeles high-rise who is out on strike. It portrays dignified people, working hard to make their way into society and fighting for their rights.

Demi, **One Grain of Rice: A Mathematical Folktale** (Scholastic Press, 1997). A village girl saves her people from starvation by tricking the raja with a mathematical equation. A powerful tale of multiplication and justice.

Judith Ennew, **Exploitation of Children** (Raintree Steck-Vaughn, 1997). An internationalist perspective on the conditions and types of child exploitation and efforts of people organizing against it. A lot of math, with great potential for use by a creative teacher.

Susan Kuklin, **Iqbal Masih and the Crusaders Against Child Slavery** (Henry Holt and Company, 1998). An excellent upper elementary biography that sets the short life of Iqbal Masih in the context of the historic struggle against child labor.

David Schwartz, **How Much Is a Million?** (HarperTrophy, 1993). A clear graphic description to help students (and teachers!) conceptualize a million and a billion.

David Smith, **If the World Were a Village: A Book About the World's People**, illus. Shelagh Armstrong (Kids Can Press, 2002). A beautifully illustrated explanation of the conditions of the world's people. Makes such topics as language, religion, health, and hunger accessible. A powerful tool for social-justice-minded teachers, particularly teachers of math.

Sources for Posters, Maps, and Additional Resources

Maps to Help People Think

www.odt.org
ODT, Inc.
P.O. Box 134, Amherst, MA 0100,
800-736-1293
odtstore@aol.com
Both the Peters Projection Map and the "What's Up? South!" world map challenge people to think differently about the world. The Peters Projection map accurately presents the area of all countries and explains how the commonly used Mercator projection distorts the sizes of continents, making Europe appear much larger than it is. The teaching guide *A New View of the World: Handbook to the Peters Projection World Map,* by Ward L. Kaiser, is also available here.

Northern Sun Merchandising

www2.northernsun.com
2916 E. Lake St., Minneapolis, MN 55406
800-258-8579, fax: 612-729-0149
nsm@scc.net
A distributor of valuable resources on environmental, gay/lesbian, multicultural, and feminist themes. Offers a particularly impressive collection of beautiful, classroom-friendly posters.

Syracuse Cultural Workers

www.syrculturalworkers.com
P.O. Box 6367, Syracuse, NY 13217
315-474-1132, fax (toll-free): 877-265-539
scw@syrculturalworkers.com
A long-time distributor of multicultural and social-justice resources, including the Peace Calendar, which should adorn all classrooms.

Teaching for Change

www.teachingforchange.org
P.O. Box 73038, Washington, DC 20056-3038
800-763-9131, fax: 202-238-0109
info@teachingforchange.org
Distributor and publisher of quality multicultural and social justice teaching materials, such as *Putting the Movement Back into Civil Rights Teaching* and *Beyond Heroes and Holidays.* This organization's catalog is the single best source for materials to help teachers to explore and teach about social justice issues.

"Videos with a Global Conscience" in *Rethinking Globalization: Teaching for Justice in an Unjust World.*

Rethinking Schools
www.rethinkingschools.org/publication/rg
1001 E. Keefe Ave., Milwaukee WI 53212
800-469-6192
Bill Bigelow describes in detail 40 videos, including *The Business of Hunger, Global Village or Global Pillage,* and *Zoned for Slavery,* giving useful teaching ideas. Many of the videos are particularly good for teaching social justice mathematics.

About the Editors

ERIC (RICO) GUTSTEIN (gutsteinrico@earthlink.net) teaches mathematics education at the University of Illinois-Chicago. He taught middle school mathematics for several years in a Chicago public school and now works with the first Chicago public high school for social justice, which opened in the fall of 2005. He is the author of *Reading and Writing the World with Mathematics* (Routledge, 2006) and is a co-founder of Teachers for Social Justice in Chicago. He is a frequent contributor to *Rethinking Schools* magazine.

BOB PETERSON (repmilw@aol.com) teaches fifth grade at La Escuela Fratney, a two-way bilingual public school in Milwaukee, Wis. He is a founding editor of *Rethinking Schools* and is a frequent writer and speaker. He co-edited (with Bill Bigelow) *Rethinking Globalization: Teaching for Justice in an Unjust World and Rethinking Columbus: The Next 500 Years* and (with Michael Charney) *Transforming Teacher Unions: Fighting for Better Schools and Social Justice.* In 1995 he was selected as Wisconsin Elementary Teacher of the Year.

Contributors

ROBERT BERKMAN has been teaching mathematics in the New York City area for the past 20 years.

BILL BIGELOW teaches at Franklin High School in Portland, Ore. With Bob Peterson he co-edited *Rethinking Globalization: Teaching for Justice in an Unjust World.* He is an editor of *Rethinking Schools* magazine.

ANDREW BRANTLINGER taught secondary mathematics in Chicago and is currently a graduate student at Northwestern University.

INDIGO ESMONDE is a graduate student in education at the University of California, Berkeley.

SAGE FORBES-GRAY is in her second year teaching seventh-grade math in a public school in Brooklyn, N.Y.

MARILYN FRANKENSTEIN teaches Quantitative Reasoning and Arguments at the College of Public and Community Service, University of Massachusetts/Boston. She writes about criticalmathematics and ethnomathematics education. Her most recent political activity was as one of the organizers for the Boston Social Forum.

SUSAN HERSH has taught in elementary and middle school classrooms in the United States and abroad for 10 years. Currently, she is a seventh-grade social studies teacher at The Preuss School-UCSD, a charter school for college-bound, low-income youth.

POLLY KELLOGG teaches in the Department of Human Relations and Multicultural Education at St. Cloud State University in Minnesota.

STEPHANIE KEMPF is a teacher and writer in New York City.

MICHAEL LANGYEL has taught high school mathematics for 28 years in the Milwaukee Public Schools.

DAVID LEVINE is a co-founder of Rethinking Schools and currently serves as a Rethinking Schools editorial associate.

LARRY MILLER is a high school teacher in the Milwaukee Public Schools, an editor of Rethinking Schools, and part of the team starting W.E.B. Du Bois High School in Milwaukee.

SWAPNA MUKHOPADHYAY is a mathematics educator who teaches mathematics as a cultural construction. She is an associate professor in the Graduate School of Education at Portland State University.

LUIS ORTIZ-FRANCO is a professor of mathematics at Chapman/University in Orange, Calif., and the past second vice president of the International Study Group on Ethnomathematics (ISGEm). He was a member of the staff of the United Farm Workers where, in collaboration with Cesar Chavez, he used his mathematics skills to secure just and fair labor contracts for farm workers.

JESSICA QUINDEL teaches at the Interactive Mathematics Program in Communication Arts and Sciences, a small school within Berkeley High School in Berkeley, Calif.

ADAM RENNER is a former high school math teacher who now teaches in the School of Education at Bellarmine University in Louisville, Ky. He instructs courses on research, social difference, service-learning, and social justice.

LAURIE RUBEL was a high school mathematics teacher for nine years and currently teaches at Brooklyn College of the City University of New York.

MEGAN STAPLES is an assistant professor of mathematics education at Purdue University.

LARRY STEELE teaches History and Business Education in the Academy of Finance, a small learning community at Franklin High School in Seattle.

BEATRIZ FONT STRAWHUN is a Research Associate and Adjunct Lec-turer in the School of Education at The University of Michigan. She has taught middle school mathematics in New York City and Austin and San Antonio, Tex.

WILLIAM F. TATE is Chair and Professor of Education and Applied Statistics at Washington University.

ERIN E. TURNER, a former bilingual elementary teacher, is currently an assistant professor of mathematics education at the University of Arizona. Her research and teaching interests include issues of equity and diversity in mathematics education.

CLAUDIA ZASLAVSKY was a noted teacher of mathematics and author of 13 books and many articles dealing with multicultural perspectives and equity issues in mathematics education. Her last book was for children: *More Math Games and Activities from Around the World* (Chicago Review Press, 2003).

Index

h

Harvey, Andrea: 96
Harvey, William: 37
Hersh, Susan: 64, 175
Hindu-Arabic number system: 21
HIV: vi, 107–108

i

In These Times: 24
Income Distribution (graph): 105
Indonesia: 62
Interactive Mathematics Program (curriculum): 5, 176
International Labor Organization: 161
Investigations (curriculum): 4
Iraq, war in: 20, 29–30, 77, 80, 120, 121, 165

j

Jaimes, Annette: 27
Joseph, George: 34, 169
Journal of Negro Education: 44

k

Keller, Helen: 25
Kellogg, Polly: 135
Kempf, Stephanie: 61, 74, 170, 175

l

laborers: 51, 62–63, 109–110, 117, 162
Ladson-Billings, Gloria: 3, 37, 116
Langyel, Michael: 68, 175
Latinos: 13, 16–17, 47–52, 70–75, 91, 116, 117
Lerman, Steve: 22
Levine, David: 90, 175
libraries: 149
Loewen, James: 142, 145
Los Angeles: 71, 97–100, 101–102
lotteries: 150–151

m

Macedo, Donald: 24
maps: 41, 111–116
Mason, Sandra: 3 8
Mathematics in Context (curriculum): 4, 112
Mayan mathematics/numbers: 70–75
McDonald's: 74
McIntosh, Peggy: 21
Mead, Margaret: 37
MegaMillions (data table): 150
Mehring, Mary Lou: 96
Mercator map: 26, 111–116, 119, 171, 172
Mexico: 113
military spending: 76–78
Military vs. Social Spending (graph): 79
Miller, Larry: vii, 41, 175
Moses, Bob: 3, 90–96, 165, 167
Mukhopadhyay, Swapna: 122, 176
multicultural education: 43–44, 70–75, 124–129

n

National Assessment of Educational Progress: 33
National Council of Teachers of Mathematics: 4–5, 10, 33, 62, 92, 112, 124, 165, 167
National Labor Committee: 56–57, 161–162, 165
National Priorities Project: 29–30, 78, 165
Native Americans: 27
Navarro, Vincente: 27
New American Century: 78
Nicaraguan election: 22
Nicaraguan factory: 56–57
No-TV Week: 14

o

Ortiz-Franco, Luis: 70, 176

p

PA announcements: 130
Peters Projection map: 26, 111–116, 171, 172
Peterson, Bob: 9–15, 64–67, 76–78, 109–110, 138–139, 140–147, 149, 160–161, 170
Polyp: 12, 59, 142

New from RETHINKINGSCHOOLS

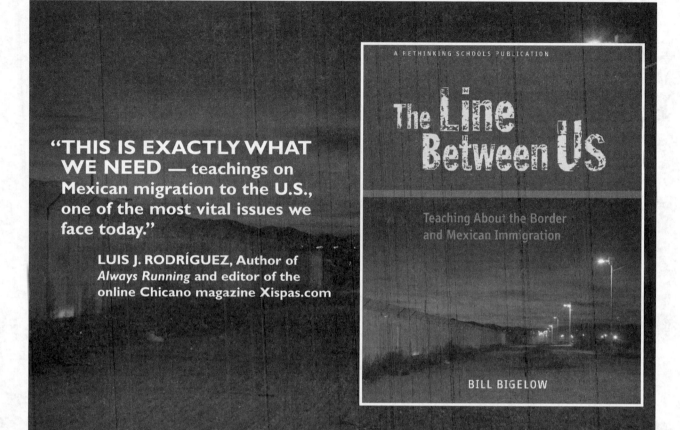

"THIS IS EXACTLY WHAT WE NEED — teachings on Mexican migration to the U.S., one of the most vital issues we face today."

LUIS J. RODRÍGUEZ, Author of *Always Running* and editor of the online Chicano magazine Xispas.com

THE LINE BETWEEN US
Teaching About the Border and Mexican Immigration
BY BILL BIGELOW

Using stories, historical narrative, role plays, poetry, and video, veteran teacher Bill Bigelow shows how he approaches immigration and border issues in his classroom. *The Line Between Us* is about imaginative teaching that gets students to care about the world.

2006 • Paperback • 160 pages
ISBN-10: 0-942961-31-5 • ISBN-13: 978-0-942961-31-7

Only $16.95! Plus shipping and handling*

"A WONDERFUL BOOK on critical teaching, filled with examples, resources, and how to transform theory into practice."

IRA SHOR, City University of NY Graduate School; author (with Paulo Freire) of *A Pedagogy for Liberation*

■ ORDER ONLINE: www.rethinkingschools.org
■ CALL TOLL-FREE: 1-800-669-4192 ■ OR USE THE ORDER FORM IN BACK

RESOURCES FROM RETHINKING SCHOOLS

RETHINKING SCHOOLS MAGAZINE

Rethinking Schools is a quarterly magazine written by teachers, parents, and education activists—people who understand the daily realities of reforming our schools. No other publication so successfully combines theory and practice while linking classroom issues to broader policy concerns.

Three years: $39.95 (Save $19.45!)

Two years: $29.95 (Save $9.65!)

One year: $17.95

Subscriptions to Canada and Mexico add $5 per year.
Other international subscriptions add $10 per year.

THE LINE BETWEEN US
Teaching About the Border and Mexican Immigration
By Bill Bigelow

Using stories, historical narrative, role plays, and videos, veteran teacher Bill Bigelow shows how he approaches immigration and border issues in his classroom.

Paperback • 160 pages
ISBN-10: 0-942961-31-5
ISBN-13: 978-0-942961-31-7
ONLY $16.95!*

RETHINKING MATHEMATICS
Teaching Social Justice by the Numbers
Edited by Eric Gutstein and Bob Peterson

Shows how to weave social justice issues throughout the mathematics curriculum and how to integrate mathematics into other curricular areas.

Paperback • 180 pages • ISBN 942961-54-4
ONLY $16.95!*

RETHINKING GLOBALIZATION
Teaching for Justice in an Unjust World
Edited by Bill Bigelow and Bob Peterson

"A treasury of ideas and information," according to historian Howard Zinn. Includes role plays, interviews, poetry, stories, background readings, and hands-on teaching tools.

Paperback • 400 pages • ISBN 0-942961-28-5
ONLY $18.95!*

THE NEW TEACHER BOOK
Finding Purpose, Balance, and Hope During Your First Years in the Classroom

Offers new teachers practical guidance on how to effectively navigate the school system and form rewarding relationships with colleagues, students, and families from all cultures and backgrounds.

Paperback • 248 pages • ISBN 942961-45-5
ONLY $14.95!*

RETHINKING COLUMBUS
The Next 500 Years

Includes more than 80 essays, poems, historical vignettes, and lesson plans that re-evaluate the legacy of Columbus. Packed with useful teaching ideas for kindergarten through college.

Paperback • 192 pages • ISBN 0-942961-20-X
ONLY $12.95!*

RETHINKING SCHOOL REFORM
Views from the Classroom
Edited by Stan Karp and Linda Christensen

A collection of the best writing from *Rethinking Schools* magazine on a broad range of pressing issues, including school vouchers and funding, multiculturalism, standards and testing, teacher unions, bilingual education, and federal education policy.

Paperback • 350 pages • ISBN 0-942961-29-3
ONLY $16.95!*

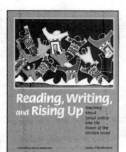

READING, WRITING, AND RISING UP
Teaching About Social Justice and the Power of the Written Word
By Linda Christensen
Essays, lesson plans, and a remarkable collection of student writing, with an unwavering focus on language arts teaching for justice.

Paperback • 196 pages • ISBN 0-942961-25-0
ONLY $12.95!*

RETHINKING OUR CLASSROOMS, Volume 2
This companion volume to the first *Rethinking Our Classrooms* is packed with more great articles about teaching for social justice, as well as curriculum ideas, lesson plans, and resources.

Paperback • 240 pages • ISBN 0-942961-27-7
ONLY $12.95!*

RETHINKING OUR CLASSROOMS, Volume 1
Teaching for Equity and Justice
Creative teaching ideas, compelling narratives, and hands-on examples of how teachers can promote values of community, justice, and equality— and build academic skills.

Paperback • 216 pages • ISBN 0-942961-18-8
ONLY $12.95!*

RETHINKING OUR CLASSROOMS, Volumes 1 and 2 Combined Set
Widely used in teacher education and staff development programs.

ISBN 0-942962-36-6
ONLY $19.95 per set!*
Save 20%!

* Plus shipping and handling.
See order form below for details.

FOUR EASY WAYS TO ORDER

❶ **CALL TOLL-FREE:** 1-800-669-4192 8am-9pm (ET) M-F ❷ **SECURE ONLINE ORDERING:** www.rethinkingschools.org
❸ **FAX ORDER FORM TO:** 802-864-7626 ❹ **MAIL ORDER FORM TO:** Rethinking Schools, P.O. Box 2222, Williston, VT 05495
MASTERCARD, VISA, AND PURCHASE ORDERS ACCEPTED

Name _____

Organization _____

Address _____

City/State/Zip _____

Phone _____

Email _____

METHOD OF PAYMENT
☐ Check or money order made out to Rethinking Schools
☐ Purchase order
☐ MasterCard　　☐ Visa

Credit Card No. _____

Exp. Date _____

Authorized Signature _____

QUANTITY	TITLE/ITEM	UNIT PRICE	TOTAL
		$	$

Subtotal	$
Shipping*	$
Donation	$
TOTAL	$

MAIL TO: Rethinking Schools, P.O. Box 2222, Williston, VT 05495

FAX TO: 802-864-7626

CALL 1-800-669-4192 FOR A FREE CATALOG OF ALL OUR MATERIALS

* U.S. shipping and handling costs are 15% of the total (minimum charge of $4.00.) Canadian shipping and handling costs are 25% of the total (minimum charge of $5.00). Subscriptions already include shipping costs. Payment in U.S. dollars.

2BRM

Resources from RETHINKING SCHOOLS

"RETHINKING SCHOOLS
offers some of the freshest, most authentic insights into schools, teaching, and students' lives of any publication available. I read it because it tells it like it is."

LINDA DARLING-HAMMOND, professor of teaching and teacher education, Stanford University

Rethinking Schools is an independent, quarterly magazine committed to equity and social justice. It is written by teachers, parents, and educational activists — people who understand the day-to-day realities of reforming our schools. It asks questions, but also looks for answers. It is grounded in classroom practice, but also analyzes broader policy issues.

Rethinking Schools has one overriding concern: to help transform our schools so they can provide a quality education for all children. Every issue includes:

- Innovative teaching ideas
- Listings of valuable resources
- Analyses of important policy issues such as the No Child Left Behind or the charter school movement.

Subscribe today and receive the country's leading grassroots education magazine.

SUBSCRIPTION RATES

$29.95 Two-year subscription *(save $9.45 off the cover price!)*
$17.95 One-year subscription
$5 Sample copy

SPECIAL BULK SUBSCRIPTION RATES are available at reduced rates
— ideal for schools of education, unions, parent groups, and classroom teachers!
Contact our office at 414-964-9646 or rsmike@execpc.com for more information.

■ ORDER ONLINE: www.rethinkingschools.org
■ CALL TOLL-FREE: 1-800-669-4192   ■ OR USE THE ORDER FORM IN BACK